Smart on a Shoestring: Redefining Smart Cities for Developing Nations

Authors

Professor Dr. Seyed Mohammadreza Ghadiri

Dr. Nguyen Thi Thanh Huong

Associate Professor Dr. Yekini Ojenya Salawu

Associate Professor Dr. Fatai A. Yakeen

Associate Professor Dr. Giwa O. Mojeed
Mr. Attah Kumah Mensah

ISBN: 978-1-962443-20-3

THIS BOOK OFFERS AN OVERVIEW OF SMART CITIES IN THE CONTEXT OF DEVELOPING NATIONS. WE SHOULD OBTAIN A SMART MINDSET FROM THE PLANNING AND BUDGETING TO THE DEPLOYMENT STAGE. THE SMART CITY EXPERIENCE FROM THE DEVELOPED COUNTRIES CAN BE LEARNT AND WELL ADAPTED BY DEVELOPING COUNTRIES WITH A PRACTICAL APPROACH. SMART CITIES ARE APPLIED IN NOT JUST TECHNOLOGY PERSPECTIVE, THEY SHOULD BE ASSOCIATED WITH COST-EFFECTIVE, SUSTAINTABLE, RESILIENT, AND INCLUSIVE DEFINITION. THIS BOOK IS IDEALLY SUITED FOR URBAN PLANNERS, POLICY-MAKERS, SMART CITY DESIGNERS, AND CITY LEADERS.

THE AUTHORS DEDICATE THIS WORK TO THEIR BELOVED FAMILIES AND THOSE WHO ASSISTED IN ITS WRITING.

PREFACE

In the contemporary era, global urban development is often guided by ambitious milestones such as the visions for 2030, 2045, and beyond. Across these timelines, nations aspire to achieve "Net Zero" emissions and transition toward high-value, knowledge-based economies. Within this narrative, the concept of the smart city has emerged as a powerful symbol of progress. It is, however, frequently portrayed through a narrow lens of futuristic skylines, ubiquitous 5G connectivity, artificial intelligence, and gleaming towers of glass and steel.

Inspired by the widely publicized successes of cities in developed nations, smartness has come to be perceived as a high-tech ideal attainable only by wealthy countries. For nearly six billion people, representing approximately 75 percent of the world's population, living in 108 middle-income countries across Asia, Africa, and Latin America, this vision often feels distant and unattainable.

In many of these contexts, the pursuit of smart cities has given rise to urban fantasies that prioritize iconic skyscrapers and automation while overlooking everyday realities. Poverty overcrowded informal settlements, unreliable services, and the absence of basic infrastructure continue to define daily life for millions. The result is a growing disconnect between global aspirations and local needs.

This book was written to challenge that paradigm. We contend that true urban intelligence lies not in the sophistication of the technologies a city acquires, but in the wisdom with which those technologies are applied to improve the lives of its people. A genuinely smart city, as we define it, is not necessarily the most technologically advanced. Rather, it is one that is cost-effective, sustainable, resilient, and inclusive. We argue for a shift toward contextual smartness, an approach in which success is measured by how effectively a city uses its existing resources to address real problems such as waste management, traffic congestion, flooding, and housing shortages, even under tight fiscal constraints.

The evidence explored in this book underscores a fundamental truth: technology is only a tool. Without a clear purpose, sensitivity to local context, and sufficient human and institutional capacity, even the most advanced systems are likely to fail. Throughout these chapters, we move beyond the "tech trap," the seductive yet flawed assumption that advanced technology is a prerequisite for smartness.

By examining stalled or struggling initiatives such as Konza Technopolis in Kenya, Forest City in Malaysia, and aspects of the Kigali Smart City vision, we demonstrate how high-cost infrastructure, when detached from local needs, maintenance realities, and inclusive governance, can result in impressive but underutilized urban landscapes. In contrast, this book champions frugal innovation and low-tech, high-impact solutions.

We draw attention to strengths that already exist in many developing cities, including dense social networks, widespread mobile phone usage, and a large, technologically adept youth population. From the culture of Jugaad innovation in India, to the digital mapping of Nairobi's matatu routes, and to SMS-based traffic alerts in Lagos, the case studies presented here show that meaningful urban transformation can be achieved through simple, locally grounded, and affordable tools.

At the heart of this work is the conviction that smart cities must be citizen-centric and must place people at the center of urban policy. We explore, in particular, the critical issue of housing affordability, arguing that a city cannot truly be considered smart if its residents cannot afford basic shelter.

To support practical implementation, we present a four-stage framework to help urban planners and policymakers assess their city's current conditions, articulate a realistic vision, achieve fiscal sustainability, and monitor progress over time. This framework is illustrated through a proposed model for Lokoja, Nigeria, which leverages the city's unique geography and historical significance to foster smart tourism and urban resilience.

Intended for urban planners, policymakers, city leaders, and smart city designers, this book serves both as a critique of prevailing trends and as a pragmatic roadmap for the future. The authors are an interdisciplinary team drawing on expertise in transportation, logistics, urban planning, and the social sciences. Together, they combine academic insight with real-world experience to offer grounded and actionable perspectives.

We dedicate this work to our families and to the many individuals whose support and encouragement made this book possible. Above all, we invite you to join us in rethinking the urban future, one in which smart choices, human ingenuity, and community engagement form the foundations of cities that are equitable, resilient, and truly sustainable.

CONTENTS

CHAPTER 1: What Makes a City "Smart"? An Introduction

CHAPTER 4: A Framework for Sustainable and Cost-Effective Smart Cities

CHAPTER 5: Institutional Capacity: Resilience Building

CHAPTER 1
What Makes a City "Smart"? An Introduction

1.1 What is a Smart City?

A Smart City is a modern approach to urban management that uses technology, data, and innovation to enhance residents' quality of life. The goal of a Smart City is to improve how a city works, including traffic flow, energy use, air quality, and public safety, by using digital tools. According to Albino, Berardi, and Dangelico (2015), Smart Cities encompass more than technology. They also involve policies, people, and the environment.

The aim is to make cities more sustainable, efficient, and habitable for everyone. The Smart City concept has evolved significantly over time. In the past, city planning focused primarily on essential physical infrastructure such as roads and bridges. Today, Smart Cities use digital technologies, including the internet, sensors, and artificial intelligence, to tackle urban problems such as traffic congestion, crime, pollution, and waste management inefficiencies. As explained by Caragliu, Del Bo, and Nijkamp (2011), a Smart City uses information and communication technologies (ICT) to enhance economic development, improve quality of life, and mitigate environmental impact.

In simple terms, a Smart City connects things like public transport, streetlights, water systems, and even buildings to a central system using digital technology. This enables city managers to know what is happening in the city in real time and react quickly when necessary. Additionally, it allows residents to use city services more easily, such as paying bills online, reporting faults through apps, or checking traffic conditions before leaving the house.

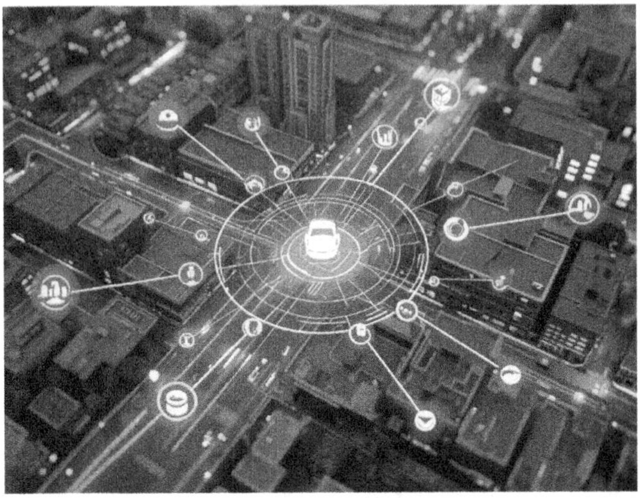

Figure 1.1: Simulation of a Smart City

1.2 Different Dimensions of a Smart City

A Smart City is more than a collection of high-tech devices; it is a complex ecosystem of interconnected systems, working seamlessly together to enhance the lives of its inhabitants. Albino et al. (2015) outline six dimensions: smart economy, smart people, smart governance, smart mobility, smart environment, and smart living.

1. *Smart Economy*: This involves using digital technologies to boost the city's economy. It involves supporting businesses that use technology and helping people acquire digital skills.
2. *Smart People*: Citizens play a crucial role in Smart Cities. People need to be educated, involved in city planning, and open to using new technologies.
3. *Smart Governance*: This refers to governments using technology to be efficient, more transparent, and connected to citizens. It includes digital documents, online voting, and open data portals.

4. *Smart Mobility*: Transport in Smart Cities should be safe, fast, and eco-friendly. This includes smart traffic lights, electric buses, and bike-sharing systems.
5. *Smart Environment*: The primary goal of this section is to protect natural resources. Smart Cities utilize sensors to monitor air and water quality and employ technology to manage waste and optimize energy consumption.
6. *Smart Living*: The goal of Smart Cities is to improve safety, wellbeing, health, housing, and culture. This could involve public safety apps, telemedicine, and online education.

According to Nam and Pardo (2011), smart cities also involve innovation in city management and policy. It is not enough to have advanced technologies; leaders must also know how to use them effectively and develop policies that support change.

1.3 Smart City Success Stories: Singapore and Barcelona

Two well-known examples of Smart Cities are Singapore and Barcelona. These cities have been recognized for the ways in which they use technology to solve problems and enhance the quality of life for their citizens. Singapore, in particular, is often regarded as the leading smart city in the world. The government has developed a comprehensive plan called "Smart Nation." Under this plan, Singapore employs sensors and cameras throughout the city to collect data on cleanliness, traffic, and energy use.

For instance, some sensors notify cleaners when a bin is full. The city also features smart traffic systems that reduce congestion and allow buses to move more efficiently. Health services in Singapore also use digital platforms to keep track of patients and book appointments with doctors online. Barcelona, in Spain, is another leading smart city. It has leveraged technology to improve public services while reducing costs. The city uses sensors to manage waste, parking, and street lighting. Residents can use apps to locate available parking spaces or report issues such as broken streetlights. In addition, Barcelona provides free Wi-Fi in many public spaces and makes open data available, enabling residents and businesses to develop new services. These cities demonstrate how technology, when used effectively, can make urban areas safer, cleaner, and more livable. However, such improvements come at a financial cost and rely on citizens' trust in the system.

1.4 The High-Tech Image of Smart Cities

Smart Cities are often perceived as high-tech environments filled with complex systems and expensive equipment. Common associations include big data, the Internet of Things (IoT), 5G connectivity, and Artificial Intelligence (AI). While these tools are undoubtedly valuable, they do not constitute the entirety of a smart city. Nevertheless, this perception shapes how people understand and accept smart cities.

The Internet of Things (IoT) is a system where everyday objects are linked to the internet and can transmit and receive data. For instance, streetlights that switch on when someone passes by or traffic lights that adjust based on traffic flow. AI can help cities predict traffic bottlenecks, enhance security through smart cameras, or assist in planning city growth. 5G makes it easier to transfer large volumes of data quickly, which is important for real-time services. Allam and Dhunny (2019) explain that big data and AI are crucial in assisting cities to make better decisions. When cities collect data about how people move, where pollution is high, or what services people use most, they can respond more effectively. However, they also caution that these technologies raise concerns about control and privacy. Kitchin (2015) adds that many Smart City initiatives place too much emphasis on technology and not enough on people. Sometimes, cities install high-tech systems without asking if people want or need them. In these cases, Smart Cities become tools for businesses and tech developers instead of truly helping citizens.

1.5 Problems and Criticisms of Smart Cities

Although Smart Cities may seem appealing, they are not without flaws. Researchers and experts have identified several issues and criticisms. Hollands (2008) was one of the first researchers to note that the idea of a Smart City is often more about marketing than real advancement.

Some cities call themselves smart just to attract business or tourism, even if they are not doing much for their residents. He suggests that really smart cities should concentrate on solving social and environmental issues, not just installing new technology.

Vanolo (2014) introduced the concept of the "smart mentality." He argues that the idea of a Smart City can also be used as a tool for social control. For instance, while collecting data from residents can improve services, it can also allow governments or businesses to monitor individuals too closely, potentially diminishing freedom and privacy. According to Söderström, Paasche, and Klauser (2014), Smart Cities are often promoted through powerful stories created by big corporations.

These stories can lead people to believe that technology alone will solve all problems. In reality, however, smart cities are often shaped by corporate interests, and the needs of ordinary residents are sometimes overlooked. In certain situations, the digital divide becomes a significant concern. If a city relies too heavily on digital solutions, residents without smartphones or internet access may be left behind. These individuals can face difficulties accessing public services, applying for jobs, or participating fully in urban life, thereby exacerbating existing inequalities rather than alleviating them.

1.6 The Need for People-Centered Smart Cities

To avoid these problems, many experts believe Smart Cities should prioritize people over machines. Nam and Pardo (2011) argue that Smart Cities should be about human ingenuity as much as technology. This means involving residents in planning, listening to their needs, and assisting them with knowing how to use new systems.

Kitchin (2015) suggests that city planners should ask crucial questions before starting any Smart City project. For example: Who gains from this technology? Who might be harmed? How will privacy be protected? How can everyone participate? Albino et al. (2015) also emphasize that cities should consider performance, in addition to innovation. This entails checking if smart tools are actually improving life in the city or if they are just being used as a showpiece. The most successful Smart Cities are those that use technology wisely, engage people in decision-making, and prioritize fairness and sustainability.

It is also important to train people to use digital tools and ensure everyone, including older people and the poor, can enjoy the benefits of smart living. It can be concluded that the concept of a Smart City is both exciting and complex. Fundamentally, a smart city uses digital technology to enhance urban life, focusing on areas such as the environment, transportation, governance, and the economy. However, smart cities are not merely defined by the presence of the latest technology; they are about improving lives in a fair, efficient, and sustainable manner.

Examples from cities like Singapore and Barcelona demonstrate that smart cities can succeed when technology is paired with careful planning and active citizen engagement. At the same time, the perception of smart cities as high-tech and expensive can make them seem out of reach for many developing countries, while also raising important questions about privacy, control, and who benefits most.

Researchers like Hollands, Vanolo, and Kitchin remind us that Smart Cities should be more than just instruments for profit or power. They should be places where technology benefits people, not the other way around. In the end, the smartest cities will be those that listen to their residents, protect their rights, and use technology not just to impress but to include, inspire and improve.

1.7 Are Smart Cities Only for Wealthy Nations?

When we hear the term "Smart City," we often imagine a modern, gleaming metropolis filled with advanced technology—high-speed internet, driverless cars, robots, and even self-cleaning windows. These images usually originate from affluent countries such as Singapore, the United States, or certain European nations. Consequently, many people assume that only wealthy countries can develop smart cities. But is this assumption accurate? Are smart cities truly limited to prosperous nations, or can low- and middle-income countries build them as well?

Figure 1.2: The Landscape of Smart City

In reality, the concept of a Smart City has become so closely associated with wealth, advanced technology, and futuristic visions that it often overlooks a large portion of the world's developing nations. Countries in Asia, Africa, and Latin America are often excluded from this vision, even though they are home to millions of urban residents who also seek clean streets, efficient transportation, and safe housing. This section examines the idea that Smart Cities are not exclusively for wealthy nations. It explores how developing countries are often left out of Smart City conversation and clarifies common misconceptions, such as the belief that a Smart City must rely on 5G connectivity, artificial intelligence (AI), or costly technology. Ultimately, it demonstrates that the true measure of a city's "smartness" lies not in its wealth, but in the creativity and ingenuity of its people.

1.7.1 The Narrow View of Smart Cities

The most prevalent concepts of a Smart City often originate from large corporations and high-tech developers, who tend to depict smart cities as highly connected and equipped with advanced digital tools. Pereira et al. (2018) observe that much of the discourse around Smart Cities emphasizes automation, data, and technology. While these tools can certainly enhance urban services, they are also costly, meaning that only wealthy cities can typically afford to implement them.

This presents a significant challenge for cities in the Global South, including those in South Asia, Africa, and Latin America. These cities frequently face a range of difficulties, such as inadequate infrastructure, overcrowded slums, and limited public funding. Yet when Smart City models are designed with technology as the primary focus, these cities are often excluded. Watson (2014) discusses the rise of "urban fantasies" in Africa.

She explains that a lot of African governments now envision building future cities filled with skyscrapers and smart technology. But in most cases, these dreams are not based on the real needs of the residents. Instead, they are copied from rich countries and do not consider local challenges like poverty or lack of basic amenities. These "fantasies" may bring foreign investment, but they ignore the daily hardships of the majority of people. Similarly, Datta (2015) discusses India's plan to create 100 Smart Cities. She refers to this project as "100 utopias" places imagined as perfect, high-tech spaces. However, these plans often exclude the urban poor and prioritize fancy housing and private transport instead of things like public toilets and clean water.

This shows that when Smart City plans follow the dreams of the rich, they can make life more difficult for the poor. While these challenges highlight the limitations of copying high-tech models from the Global North, they also open opportunities for cities in the Global South to create their own definitions of smartness based on local ingenuity and participation.

1.7.2 Humility and Local Solutions

According to Odendaal (2020) Smart Cities in the Global South need a different strategy that focuses on humility and local expertise. She argues that instead of copying Europe or North America, cities in the South should come up with smart solutions that work for their unique situations. This means examining what local communities are already doing and building on those ideas. For example, in many African cities, people use mobile phones in creative ways to find transport, manage their businesses, or coordinate neighborhood security. These are smart practices, even if they do not use the latest worldwide technology.

Odendaal advocates for "small-scale, people-led innovation" instead of large, top-down projects designed by multinational corporations. Brelsford et al. (2017) also argue that urban development should begin with actual neighborhood patterns, especially in places where informal settlements are prevalent. These are areas often called "slums," where millions of people reside. Instead of destroying these areas to build new "smart" cities, governments should study how they develop and find ways to improve them using low-cost but effective solutions. This shows that smartness can be simple. It can mean cleaner environments, better planning, and more inclusive services even without advanced technology.

1.7.3 Common Myths About Smart Cities

Let's examine some of the common misconceptions regarding Smart Cities:

Myth 1: Smart Cities Must Have High-Tech Tools Like AI and 5G

Many people believe that Smart Cities must be full of AI (artificial intelligence), 5G internet, and other expensive devices. While these tools can be useful, they are not the only way to be smart. According to Kitchin (2016), focusing too much on technology leads cities to forget the moral, political, and social parts of urban life. In fact, many wealthy cities have failed in their Smart City plans because they have only focused on buying the latest gadgets. These tools are costly, need frequent updates, and often don't reflect the real needs of the populace. On the other hand, cities in developing countries have found simple but smart ways to solve problems using local resources, community involvement, and simple digital tools.

Myth 2: Smart Cities Are Only Found in Rich Countries

This is untrue. Many developing cities are already using smart concepts. For example, Nairobi in Kenya has used mobile money (like M-Pesa) to change the way people conduct business and pay their bills. In Latin America, cities like Medellín in Colombia have enhanced transportation and public safety using community involvement and data collection, not expensive technologies. Zubizarreta et al. (2015) argue that a Smart City should be one that uses all its resources, human, social, and cultural, not just technical. A city becomes smart when it improves the life of its people in smart ways, regardless of how rich it is.

Myth 3: Smart Cities Are Built by Businesses

Many Smart City initiatives are led by big technology companies. These companies design systems and market them to governments. Hollands (2020) warns that when companies take control of smart city development, the emphasis shifts from public benefit to private profit. Instead of benefiting all citizens, smart systems might only benefit those who can afford them. Cardullo and Kitchin (2019) also warn against the concept of "smart citizenship," which tends to include only those who use digital services. This approach excludes poorer or older residents who may lack smartphones or internet access. They argue that cities must ensure the participation of all citizens in decision-making, not just the technologically savvy.

1.7.4 Rethinking What It Means to Be Smart

It is time to reconsider what a Smart City means. Being smart is not only about having money or high-tech gadgets. It's about thinking creatively and solving real problems with the available resources. Barcelona, for instance, is often cited as a successful Smart City. But as Bakıcı et al. (2013) explain, their achievement came not just from buying new tech, but from planning carefully and involving the residents.

They reduced waste, improved transport, and supported local businesses with both technology and good governance. This demonstrates that a city can be smart if it listens to its residents, manages resources efficiently, and provides equal opportunities for all. Cities in the Global South have already shown impressive examples of smart practices without relying on large budgets.

1.7.5 Smart Governance Over Smart Gadgets

Smart governance means using the best ideas and resources to make decisions, provide services, and involve citizens. According to Pereira et al. (2018), good governance is central to Smart Cities. This includes being transparent, listening to public feedback, and long-term planning. For many cities in developing countries, smart governance can be more beneficial than smart gadgets. This could mean creating local apps to report broken roads or power cuts, or using simple mobile surveys to ask people about water problems. These are low-cost but high-impact solutions. Odendaal (2020) explains that when city leaders are modest, willing to work with communities, and open to learning, even low-income cities can become smart in ways that matter most to people.

1.7.6 Who Is Left Behind?

One of the main issues with the current Smart City narrative is exclusion. People who live in informal areas, lack internet access, or are not tech-savvy are often excluded. They may not be able to pay for data, access smart services, or even understand how the systems operate. Watson (2014) points out that in many African countries, Smart City plans have resulted in the eviction of poor people from valuable property to build smart infrastructure and luxury buildings.

This demonstrates how Smart City concepts can actually increase inequality if not handled properly. To address this, Smart City planners must always ask: Who will gain from this plan? Who might get hurt? How can we ensure everyone can live in the city?

1.7.7 Conclusion: Smartness Isn't About Wealth, It's About Ingenuity

The question, then, is: are Smart Cities only for wealthy nations? The answer is no. While affluent countries may have the resources to acquire advanced technologies, true smartness lies in understanding local challenges and addressing them through innovative and inclusive solutions. Developing countries may lack the resources for robots or smart highways, but they possess something even more powerful: people who are determined, creative, and resourceful. Whether it is using bicycles for transportation, mobile phones for banking, or community meetings for urban planning, these are smart actions that enhance city life. It is time to move beyond the notion that smart cities must resemble science fiction. True smartness lies in being wise, just, and creative. It means prioritizing people and making the most available resources to improve everyday life.

1.8 Urban Realities in Developing Countries: Challenges and Constraints

Urban areas in developing countries are expanding very fast. Many people are relocating from rural villages to big cities in search of jobs, health care, better education, and a better quality of life. But this rapid migration of people is creating many problems. Cities are expanding faster than their governments can plan or control, and this causes poor housing, overcrowding, social inequality, and a lack of basic services.

This section examines the challenges currently faced by cities in developing countries. It addresses issues such as rapid urban growth, inadequate access to clean water, overcrowded slums, limited government funding, poor transportation, and various social challenges. These difficulties highlight why smart city solutions are not merely beneficial, but essential. The discussion also emphasizes that cities must be smart not only through technology, but through simple, locally grounded solutions that address real problems for real people.

1.8.1 Rapid Urbanization and the Urban Boom

According to the United Nations (2018), more than half of the world's population now lives in cities, with the majority of this growth occurring in developing countries in Asia, Africa, and Latin America. Cities such as Mumbai, Lagos, Dhaka, and Nairobi are expanding rapidly, as thousands of people migrate to these urban centers each day in search of work and opportunities. This phenomenon is known as urbanization. In many cases, however, cities are not prepared to accommodate such large numbers of people.

There are insufficient jobs, housing, schools, and roads. As a result, many residents construct their own homes on land that is not officially approved for housing. These areas, known as informal settlements or slums, often lack basic necessities such as clean water, electricity, and waste management. According to Glaeser (2014), this type of urban growth has both positive and negative aspects.

On the positive side, cities provide greater opportunities for employment and education. On the negative side, problems arise when city governments cannot deliver services quickly enough to keep pace with the rapidly growing population.

1.8.2 Informal Settlements: Life in the Slums

A large population in developing countries lives in slums. Davis (2006) in his book *Planet of Slums,* explains that over 1 billion people around the world live in poor conditions. Slums are areas with poor sanitation, overcrowded homes, and very little government assistance. People living in slums often face health issues because of unclean water and poor hygiene.

Children may not attend school, and adults may not find stable jobs. The houses are made of weak materials and can be damaged by heavy rain or fire. These people are often ignored by city planners and governments. Marx, Stoker, and Suri (2013) argue that slums are not merely a housing issue, but an economic one. They exist because formal housing is often too expensive, and city governments have not prioritized the provision of affordable housing. In addition, there is frequently insufficient land properly designated for new housing development.

1.8.3 Limited Resources and Poor Infrastructure

Developing cities also face a severe shortage of funding and resources. Schools, roads, hospitals, and water systems are all expensive to build and maintain. When urban populations grow rapidly, infrastructure often struggles to keep pace. Angel et al. (2011) argue that the majority of developing cities have not been properly planned for growth. They say that city planners must "make room" for new people by reserving land, building roads, and creating basic services in advance. Without such planning, cities risk becoming chaotic, experiencing traffic congestion, water shortages, and frequent power outages.

Satterthwaite (2016) found that many urban areas fail to meet even the most basic needs of their residents, such as access to clean water and adequate sanitation. In many slums, families share one toilet, or they may not have access to clean drinking water. These problems lead to sickness and death, especially for children.

1.8.4 Weak Governance and Corruption

A major cause of the poor condition of cities is bad governance. Many city governments do not have the power or the resources to make the changes needed. In some cases, leaders are corrupt or indifferent to poor communities. Rather than supporting residents of informal settlements, governments may attempt to evict them. According to Huchzermeyer (2011) in many African cities, governments used to see slums as something that needed to be destroyed. However, perspectives are changing. There is growing recognition that slum residents have a "right to the city," including the right to remain in their communities and improve their homes with support from the government. Nevertheless, granting rights is not enough. Mitlin (2018) explains how urban communities are organizing themselves to demand better services. These urban social movements use non-violent methods to cooperate with or oppose governments to bring about real change. Still, many city leaders lack planning tools, experience, or long-term vision. They often rely on international aid, and that aid may come with conditions attached or be mismanaged. Poor governance often reinforces inequality. When public resources are mismanaged or policies overlook marginalized communities, the gap between the rich and poor widens, leading to social fragmentation.

1.8.5 Social Inequality and Exclusion

Cities in developing countries are often sharply divided between rich and poor. While some neighborhoods feature shopping malls, luxury homes, and well-maintained roads, other areas of the same city suffer from open drains, informal settlements, and lack of electricity. This stark inequality is both dangerous and unjust. Poor residents frequently live far from city centers, with limited access to reliable transportation and employment opportunities. They may also face discrimination based on ethnicity, income, or migrant status. This kind of inequality makes cities less productive and more likely to suffer from unrest, crime, or protests. It also keeps large groups of people trapped in poverty with little opportunity to improve their lives.

1.8.6 The Importance of Smart City Approaches in Developing Contexts

In many wealthy countries, the term "Smart City" is often associated with cutting-edge technology used to enhance urban life. In developing countries, however, smart solutions do not need to be high-tech or costly to be effective. Smart City thinking is important for cities with limited resources as well, but "smartness" should focus on addressing real problems through creative and simple solutions. For example:

- Using mobile devices/phones for public health updates and information.
- Installing solar lights in slums without electricity.
- Using apps to report waste problems or water leaks.
- Creating affordable public transport systems that reduce traffic congestion.

What makes a city truly smart is not the number of sensors or robots it possesses, but how effectively it uses its existing resources to benefit all residents, including the poor. This emphasis on context-sensitive, affordable solutions lays the groundwork for exploring how resource-constrained cities have developed "frugal innovations" that are both practical and transformative.

1.8.7 Applying Frugal Innovations in Urban Management

In India, cities are embracing "jugaad" innovation, a concept that emphasizes finding low-cost solutions to problems. Borkar and Pande (2022) describe how Indian cities are using basic technologies to improve urban services. For instance, some cities employ mobile apps to manage waste collection, while others use inexpensive sensors to monitor air quality. These frugal innovations demonstrate that smart solutions do not have to be costly to be effective.

1.8.8 Digital Infrastructure

Digital technologies can improve urban mobility. In Nairobi, Kenya, digital maps of informal bus routes (matatus) help people navigate the city. Graham (2020) explains how this digital infrastructure enhances transportation. By using technology, cities can provide better services even with limited resources. Digital tools can help manage traffic, plan routes, and provide real-time information to residents.

1.8.9 Making the Case for Smart Planning

Given all the challenges, poor infrastructure, rapid growth, slums, and weak governance, developing countries need smart planning more than ever. Cities need to:

- Plan for future population growth.
- Make affordable land available for housing.
- Provide basic services like water and electricity to all.
- Support local solutions that come from the community.
- Work closely with citizens to build trust.

Angel et al. (2011) suggest creating urban expansion areas where future buildings can be placed in an organized way. This avoids the dangerous and messy sprawl of informal settlements. In addition, Mitlin (2018) emphasized how communities themselves can come up with solutions if they are given a voice and resources.

1.8.10 Reimagining the Future of Urban Development

The challenges of cities in developing nations are real, but they are not impossible to solve. The key is not just machines or money. The key is strong community engagement, inclusive planning, and smart use of available resources. Even poor cities can be smart. A small public toilet system that is safe, clean, and cheap to use is smarter than a billion-dollar smart parking system in a wealthy city.

A mobile phone health alert in a slum may save more lives than a smart camera in a shopping mall. Cities must prioritize people. They should make decisions that reduce inequality, enhance quality of life, and plan thoughtfully for the future. Technology should serve as a tool to support these goals, not as a means of showing off.

Figure 1.3: With a purpose, Smart Cities should be citizen-centric

1.8.11 Conclusion: A Call for Smarter Cities with Purpose

Urban challenges in developing countries are serious and complex. Rapid urbanization, weak governance, slums, and inequality are not just problems; they are the reality for millions of people. These challenges demand smart thinking, not necessarily expensive technology, but clever, people-centered ideas.

The future of developing countries depends on how well they respond to today's problems. If cities use their resources prudently, including the poor in decision-making and planning, they can grow into fairer, healthier, and smarter places. Ultimately, smartness is not about money; it's about creativity.

References

Ahvenniemi, H., et al. (2017). *What are the differences between sustainable and Smart Cities?* Cities, 60, 234-245.

Albino, V., Berardi, U., & Dangelico, R. M. (2015). *Smart Cities: Definitions, dimensions, performance, and initiatives.* Journal of Urban Technology, 22(1), 3-21.

Allam, Z., & Dhunny, Z. A. (2019). *On big data, artificial intelligence and Smart Cities.* Cities, 89, 80-91.

Angel, S., et al. (2011). *Making room for a planet of cities.* Lincoln Institute of Land Policy.

Bakıcı, T., et al. (2013). *A Smart City initiative: The case of Barcelona.* Journal of the Knowledge Economy, 4(2), 135-148.

Borkar, S., & Pande, H. (2022). *Frugal Smart Cities: How Indian cities are using Jugaad innovation.* Technological Forecasting and Social Change, 176, 121442.

Brelsford, C., et al. (2017). *Toward cities without slums: Topology and the spatial evolution of neighborhoods.* Science Advances, 4(8).

Caragliu, A., Del Bo, C., & Nijkamp, P. (2011). *Smart Cities in Europe.* Journal of Urban Technology, 18(2), 65-82.

Cardullo, P., & Kitchin, R. (2019). *Smart urbanism and smart citizenship: The neoliberal logic of 'citizen-focused' Smart Cities in Europe.* Environment and Planning C, 37(5).

Datta, A. (2015). *A 100 Smart Cities, a 100 utopias.* Dialogues in Human Geography, 5(1), 49-53.

Davis, M. (2006). *Planet of Slums.* Verso Books.

Glaeser, E. (2014). *A world of cities: The causes and consequences of urbanization in poorer nations.* NBER Working Paper.

Graham, S. (2020). *Digital infrastructures of urban mobility: Lessons from Nairobi's digital matatus.* International Journal of Urban and Regional Research, 44(3).

Hollands, R. G. (2008). *Will the real Smart City please stand up?* City, 12(3), 303-320

Hollands, R. G. (2020). *Beyond the corporate Smart City?* Urban Studies, 57(8), 1765-1780.

Huchzermeyer, M. (2011). *Cities with 'slums': From informal settlement eradication to a right to the city in Africa.* UCT Press.

Kitchin, R. (2015). *Making sense of Smart Cities: Addressing present shortcomings.* Cambridge Journal of Regions, Economy and Society, 8(1), 131-136.

Kitchin, R. (2016). *The ethics of Smart Cities and urban science.* Philosophical Transactions A, 374(2083).

Marx, B., et al. (2013). *The economics of slums in the developing world.* Journal of Economic Perspectives, 27(4), 187-210.

Mitlin, D. (2018). *Beyond contention: Urban social movements and their multiple approaches to secure transformation.* World Development, 107, 223-232.

Nam, T., & Pardo, T. A. (2011). *Smart City as urban innovation: Focusing on management, policy, and context.* Proceedings of the 5th International Conference on Theory and Practice of Electronic Governance.

Odendaal, N. (2020). *Smart Cities in the South: A case for humility.* International Journal of Urban Sciences, 24(3), 421-436.

Pereira, G. V., et al. (2018). *Smart governance in the context of Smart Cities: A literature review.* Information Polity, 23(2), 143-162.

Ramaswami, A., et al. (2016). *Meta-principles for developing smart, sustainable, and healthy cities.* Science, 352(6288), 940-943.

Satterthwaite, D. (2016). *Missing the Millennium Development Goal targets for water and sanitation in urban areas.* Environment and Urbanization, 28(1), 99-118.

Söderström, O., Paasche, T., & Klauser, F. (2014). *Smart Cities as corporate storytelling.* City, 18(3), 307-320.

United Nations (2018). *World Urbanization Prospects.* [UN Report]

Vanolo, A. (2014). *Smart mentality: The Smart City as disciplinary strategy.* Urban Studies, 51(5), 883-898.

Watson, V. (2014). *African urban fantasies: Dreams or nightmares?* Environment and Urbanization, 26(1), 215-231.

Zubizarreta, I., et al. (2015). *Smart city concept: What it is and what it should be.* Journal of Urban Planning, 140(1).

CHAPTER 2
Debunking the Smart City Myth

2.1 Introduction

All nations envision a compelling scenario when shaping national strategies for building Smart Cities over at least five years. However, if not started immediately and with sufficient commitment, the five-year stage serves merely as the beginning and preparation for an ambitious vision. For example, the Smart City milestone by 2030, along with the vision for 2045, represents a long-term race for achievement. The years 2030 and 2045 are target years for multiple objectives and for the quantification of national initiatives, including not only Smart Cities but also Net Zero goals, the transition to high-income status, and a shift from labor-intensive production to high-value, knowledge-based industries.

Developing countries do not all start together; rather, they progress at their own pace and learn from each other within their specific economic contexts. The pursuit of Smart Cities is also aligned with the achievement of targeted economic growth, as developing countries strive to overcome the middle-income trap. Undeniably, Smart City transformation is a global trend, representing the most advanced model of urban digital transformation and a forward-looking orientation. However, Smart Cities are unavoidably challenged by the need to address multiple demanding targets simultaneously.

It is evident that smartness must be understood in relation to the socio-economic realities of developing nations. Typically, the World Bank has found that since 1990, only 34 middle-income economies have successfully transitioned to high-income status. Among these, more than one third benefited either from integration into the European Union or from oil resources. As of the final quarter of 2023, there were 108 countries classified as middle-income, each with an annual GDP per capita ranging from USD 1,136 to USD 13,845. These 108 countries are home to approximately six billion people, accounting for about 75% of the world's population. Despite generating more than 40% of global GDP, they are responsible for over 60% of global carbon emissions, and two out of every three people living in these countries experience poverty.

As a consequence, the challenges faced by these countries are considerably greater than those encountered by their predecessors in escaping the middle-income trap, including rapidly aging populations, rising protectionism in advanced economies, and the urgent need to accelerate the energy transition. The risks of the middle-income trap are widely recognized as serious obstacles (World Bank, 2024).

In such motivations, building Smart Cities is an aspiration fostered by not only developing countries' leadership but also the entire ecosystem, including policymakers, government officials, urban planners, city administrators, academics, international organizations, entrepreneurs, and innovators, in the shared efforts to improve the quality of people's lives. Many opinions suggest that when planning urban projects, developing countries should choose the Smart City model to develop, optimizing the factors of "modernity" and "synchronization" of the living environment. Developing countries are increasingly leveraging and transforming cities based on the simulation characteristics of smart cities.

They should be able to identify that it's time to seize the unique opportunities and overcome the potential challenges. This requires more time, practice, a universal approach, and a localized mindset. In this Chapter, the authors will debunk common myths and unveil the wrong prerequisites for advanced technology and costly infrastructure in a Smart City. In the context of urban trends towards "smartness", technology is not the sole factor; the definition of Smart Cities is additionally enhanced by resource efficiency, data-driven governance, and community engagement. Governments focus on fostering a citizen-centric life and diversifying the benefits for beneficiaries within the smart cities' ecosystem. In well-developed countries, Smart City concepts have been leveraged every 5 years, seen in the sustainability practices, equity, and inclusion growth.

Smart Cities aim to address the challenges and ease the burdensome consequences of urbanization, including long-term carbon emissions, traffic congestion, water, air, and noise pollution, disasters, unbalanced migration, unemployment, population aging, and issues related to intellectual property rights. However, research suggests that a "tech-first" approach remains dominant in several developing countries, despite its relatively high costs, maintenance challenges, and frequent failure to align with local needs in low-income contexts. This chapter identifies key weaknesses that have led, or may lead, to the failure of certain projects or initiatives in developing countries, synthesizes the main contributing factors, and proposes recommendations to mitigate the risks associated with such projects.

2.2 The Tech Trap: Why Gadgets Alone Don't Make a City Smart?

Smart Cities are often perceived as futuristic urban environments built on cutting-edge technology and sophisticated infrastructure. It is tempting to picture a Smart City as a glittering horizon of digital marvels. The tech-first approach to Smart City development has proven a significant trap in various contexts, particularly in developing countries. While innovations like sensor networks, AI-driven traffic management, and smart grids sound promising, their implementation often faces significant hurdles. Yet, behind the glossy promise lies a stubborn truth: technology by itself does not make a city truly *smart*. City leaders are extremely eager to invest heavily in the overpriced devices and expensive software, yet they may not be aware of the fragile context. Even more critically, without considering the local needs, city leaders take bold and overambitious decisions to import technologies from the most advanced economies in North America or Europe and deploy them in low-income localities across Asia and Africa. They have neglected the reality that true smartness lies not in the gadgets but in the wisdom of applying them. It is in listening to the city's pulse, understanding its social facts, and shaping technology as a servant, not a master.

Lastly, the dream of a tech-driven utopia risks becoming a collection of shiny ruins, echoing with the question: *Who was this built for?* For several decades, the global conversation around Smart Cities has been driven by a seductive assumption: that advanced technology is a prerequisite for achieving "smartness".

Figure 2.1: Visualizing the linkage of a technology system in a Smart City

If a municipality installs numerous sensors, constructs multiple data towers, and layers artificial intelligence throughout the city, and integrates 5G, Internet of Things (IoT), as well as Cloud Computing to reshape interaction circle among "Government -

Institutions - Entrepreneurs - Citizens", it will surely become smarter. This vision is easy to accept - a dazzling skyline of automated systems and futuristic dashboards. But beneath that promise lies a dangerous flaw. Technology, by itself, does not mean an excellent city. Technology is only a tool, and like any tool, it needs a clear purpose, the appropriate context, and human capacity to deploy and govern it effectively. When cities fail to consider an intelligent approach to purchasing both hardware and software, they underestimate the real foundations of urban improvement in city governance, community participation, and a deep understanding of local needs.

Looking at developing countries, these flaws can lead to costly assumptions or procurement decisions. Municipal leaders often procure high-tech Information and Communication Technology (ICT) projects, including AI-driven traffic lights, sensor-laden waste bins, and smart grids, without first addressing basic infrastructure needs and the gaps in local team qualifications required to moderate and maintain these systems.

The lack of skills to immediately resolve problems such as electricity outages or Internet disconnections can interrupt the entire city network. Local engineers must be able to anticipate overall network behavior and adapt using informal problem-solving skills drawn from situational street life, the daily management of half-functioning gadgets, and screens filled with error messages.

With incorrect installations, chaotic transportation, and unreliable services forming a paradox, cities are unable to convince their residents to settle down. These issues are made more severe by a lack of innovation and alignment.

Technology should serve citizens' well-being, not merely the interests of autocratic and overly decisive mayors. Thus, urban planners must balance their logical thinking and practical mindset to avoid the technology trap. A smart city journey should begin with a human-centered approach, prioritizing safety, mobility, and dignity. Next, local demand should be matched with technology, carefully considering available options, sustainability, and local support. The comprehensive "WH" question set should always be considered when leaders decide to choose a specific technology for their future smart cities:

- *What exact problems are we solving?*
- *What are our real expectations for the mid-term and long-term periods?*
- *What emerging issues of other urban zones are often tackled with this technology?*
- *What state-of-the-art solutions are we empowered to find?*
- *What devices can we buy?*
- *Why is this technology needed to improve the existing system?*
- *Who are the final beneficiaries, and have we consulted them?*
- *How to obtain and analyze the ratios/results of the referendum?*
- *Where will this technology be deployed, and does the environment support it (power, internet, security, local usage habits)?*
- *Why should citizens trust and adopt this technology, and what value will they see?*
- *Who will use, maintain, and manage this technology locally?*
- *What training, governance, and budget are needed to keep it running?*
- *What method of fund mobilization should we adopt?*
- *What are the timelines, aligned with deliverables/outputs?*
- *What happens if this technology fails, and what kind of backup plan should we adopt?*
- *Which alternative solutions (local-friendly, low-tech, or hybrid) have we compared with the previous technology?*

To exemplify the above rationale, the project "Konza Technopolis" in Kenya is a typical case illustrating delays, red tape, and funding shortfalls due to its overly ambitious and over-budget technological vision. The government of Kenya launched the project in 2011, as a flagship project of the Vision 2030 agenda. Although nine years remained until the maturation of the Master Plan, it can be evaluated as a design flaw and a Public-Private Partnership (PPP) failure, due to limited expertise during the initial planning phase and significant funding gaps.

Technically, specialists designed the project as a replica of Silicon Valley, with highly advanced, high-cost infrastructure, including smart utilities, high-tech data centers, large-scale investments in imported systems, and costly long-term maintenance requirements. However, the project lacked a foundational ecosystem to support that vision, including a skilled labor base, nearby universities, or supporting industries such as finance and engineering. There was a significant disparity between projected demand and the existing ecosystem.

Most of the surrounding businesses and workers focused on basic goods and low-skilled jobs, with limited access to high-end tech services. There were no strong venture capital networks, incubators, or clusters of tech firms to support growth. Another critical issue was the funding gap: the project required $3.97 billion, with only around 6% actually funded by the government for infrastructure development; the rest was expected from private investors to develop universities, offices, housing, and hotels.

Kenya's government and private partners could not secure the promised and allocated funding streams due to delays in land acquisition and budget execution, leading to soaring interest rates. Political pressure intensified because the government could not accurately forecast political risks, as the cutting-edge technology came with higher price tags, highlighting the mismatch between the grand vision, actual financial capacity, and the institutional governance capability required to implement the project. Finally, the project's market-first approach prioritized private capital and profit over local community needs, and government support failed to establish a solid economic or institutional base (Nduru, 2021). Another clear example is the Kigali Smart City initiative in Rwanda, which is expected to be the "Silicon Valley of Africa." To transform Rwanda into a knowledge-based economy, Kigali aimed to develop a holistic ecosystem with a critical mass of talent, R&D, and business-led innovation, thereby accelerating socio-economic impacts. This strategic project sought to position Kigali as a regional hub for ICT events and innovation centers.

Its vision was built almost entirely around technology. Kigali was a key part of the Smart Rwanda 2020 Master Plan, which was later refined into the Smart City Rwanda Master Plan and launched in 2017, reflecting Rwanda's strongly technology-driven ambition. Kigali's Smart City roadmap was impressive in scope but fell short in execution. It focused on deploying the most modern technology without the timely preparation of appropriate supporting infrastructure, local capacity, and inclusive governance frameworks. This mismatch turned many tech pilots into fragile experiments rather than sustainable improvements.

Initially, a bold vision was built for Kigali that included (1) networked connectivity through smart street lighting, citywide sensors, and ICT kiosks; (2) a central command-and-control system to monitor traffic, security, utilities, and real-time waste management; and (3) big data analytics and AI for detecting waste dumping, as well as participatory digital platforms for citizen feedback. This comprehensive vision portrayed Kigali as an innovation and digitalization hub ready for a smart future (UNHABITAT, 2025). According to the intergovernmental perspective of UN-Habitat (2025), despite such a grand vision, critical underperformance and shortfalls in key delivery areas were identified. The plan was designed to operate with a reliable power supply and high-speed Internet, while Kigali was still in the process of upgrading its major grid and broadband network. These shortfalls led to inconsistent power supply and negatively impacted project execution. Moreover, scholars frequently identified maintenance gaps and skill deficits. The budget only covered procurement and installation, without funding for long-term equipment maintenance services and periodic training for local technicians.

Last but not least, the master plan was deployed with a top-down, technology-driven approach and showed little adaptation to local needs. The community was not involved in providing feedback; thus, the project was rolled out despite low adoption levels and limited relevance to residents' daily lives. A vision remains merely a vision if it is not grounded in real-world practice.

The World Bank praised Malaysia, an Asian developing country, for driving remarkable growth that elevated the country from low- to upper-middle-income status, and for being well prepared for a high-income transition in the coming few years (World Bank, 2025).

Malaysia learned an important lesson from the low housing occupancy rate of the real estate project Forest City, a Smart City in Johor. Forest City was once marketed as a flagship model of future cities in Southeast Asia. This project received several international awards for being a smart and green urban development:

- Global Human Settlements Award on Planning and Design in 2016.
- Global Model of Green Building Industrial Park in 2017
- Global Model on City-Industry Integration in 2018
- Global Green Smart City in 2019
- 5th International Data Corporation (IDC) Smart City Asia Pacific Awards (SCAPA) in 2019 for "Smart B.I.A" system.
- Global Model of Coastal Ecological Environment Protection in 2020
- Global Model of Low-Carbon City Planning and Design in 2021

- Global Model of Green and Intelligent Construction Award in 2022.

The failures of Forest City showed that any developing country could fall into the same trap as others by investing heavily in futuristic technology and grand designs, yet failing to align them with citizens' needs, realistic urban planning, and long-term community engagement. As planned, Forest City promised a futuristic urban hub, integrated with green technology, AI-driven security, automated transport, and high-end digital infrastructure.

As a leading Chinese developer in Malaysia, Country Garden was overly confident and invested massively in luxurious technological features, while also being subject to significant policy impacts imposed by the Malaysian government on financial flows from China.

Among these was the innovative "Smart B.I.A." system, which integrated the newest technologies, such as BIM–IoT–AI, combining BIM (building information modeling), IoT sensing platforms, and AI city applications, including green drip irrigation systems, building automation systems, intelligent security systems, smart meters, emergency systems, community robots, and other integrated systems (Newswire, 2019). However, the reality turned out to be very different from this ambitious project.

Impressive technologies caused unit prices to rise dramatically, putting them beyond the reach of ordinary Malaysians. This luxury real estate segment became too unaffordable, despite the project's aim to foster a thriving smart community. As a consequence, technology-driven Smart Cities remained largely empty and unsustainable, with many towers left unoccupied.

Massive high-tech buildings and smart villas, however, served little purpose. As we can see from the three case studies above, people often believe in the myth that more gadgets automatically equal more wisdom. True cities are not made smart by machines alone; they are made smart by the thoughtful way in which technology is chosen, applied, and sustained. To address and criticize these flaws, it is essential to recognize that a real Smart City begins with people, policies, and practical solutions. Advanced technology can amplify these foundations, but it cannot replace them. If urban experts do not acknowledge these factors, the tech trap will continue to lure Smart Cities into expensive dreams that fail to meet their most realistic expectations.

2.3 Lessons from the Ruins: Failed Smart City Experiments

Many people often envision a promise of opportunity when thinking of smart cities. That perspective is not false, as Smart Cities are designed to bring communities a happier and more fulfilled life. Essentially, a Smart City adopts a more cutting-edge approach by leveraging new technologies such as 5G, the Internet of Things (IoT), cloud computing, and other core fields to change the way governments, businesses, and people interact with each other across livelihoods, environmental protection, public safety, urban services, and industrial and commercial activities, thereby improving the efficiency of urban operations and creating a better urban life for residents. Smart Cities promote sustainable living spaces, applying modern technology to connect infrastructure and urban utilities in order to improve the quality of life of residents and support the government in effectively managing urban zones (Le, 2023).

Pursuing an overambitious tech component, leaders often failed to respect the need to combine three pivotal bases: (1) Technologies, (2) Institutions, and (3) Citizens. In many developing nations, without genuine community engagement, their vision collapsed under the weight of technology deployed in Smart City projects. As a result, villas and iconic skyscrapers became empty showpieces, rather than a thriving community. A vital lesson learned from this failure is that without grounding high-tech plans in local realities, even the most advanced systems will falter.

i. Jakarta Smart Grid Pilot, Indonesia

As part of the Jakarta Smart City Initiative, the city launched a Smart Grid Program in the mid-2010s to address chronic blackouts and overloaded power lines. Jakarta leaders imported meters and control systems to install across a wide range of selected districts, with high expectations of rapid improvements in energy efficiency. In reality, the outcome was different; residents continued to complain about billing increases and distrusted the unreliable devices.

The deficiency in clear financial incentives made it harder for lower-income households to afford the devices and recognize their value. The limited consumer awareness caused hesitation in replacing a familiar system with a smart meter, which was intended to enable benefits from solar and wind energy. The cause of frequent malfunctions, long repair delays, and device downtime was that local technicians were not trained or well-prepared to operate and maintain the advanced foreign equipment.

After a few years of use, most of the pilot's smart features were offline, so the old problems remained unsolved, even as the accrued expenses were considerable. A clear lesson emerges even cutting-edge hardware can fail or remain unused without proper training, funding for maintenance services, and public understanding. Otherwise, Smart City ambitions risk becoming stranded technologies rather than driving urban transformation (Loso Judijanto, 2024).

ii. Rio de Janeiro Operations Center, Brazil

During the run-up to the 2014 World Cup and the 2016 Olympics, Rio de Janeiro invested heavily in a state-of-the-art Operations Center, which comprised multimillion-dollar facilities. This project was recognized as a key component of Brazil's high-profile smart city strategy, developed through a partnership between the Rio City Government and IBM. The project was expected to create a center that integrates utilities, police, the central fire department, the health department, and other relevant agencies, monitoring everything from weather and traffic to trash collection, electricity, water, gas, disease outbreaks, and emergencies.

In urgent situations, such as a natural disaster or security threat, a crisis room was made available for the mayor and his advisors to make executive decisions. With its core function of increasing the efficiency of service delivery and ensuring timely responses to disruptions and service failures, this Operations Center helped Rio de Janeiro become not only a safer city but also a more sustainable and inclusive hub (EPA, 2025).

Initially, the Operations Center became a showcase for Brazil's emerging Smart City project. Delegations from other countries visited and evaluated the facility, expressing admiration for its dashboards and control rooms. Scholars such as Pereira et al. (2017) noted that the center symbolized the potential for municipal governments to leverage technology to manage complex urban challenges. However, comprehensive studies revealed the project's shortcomings.

Moutinho et al. (2016) reported that the center brought unequal benefits. The wealthier districts received better and faster responses to traffic incidents and weather alerts, while service delivery for basic sanitation in informal settlements, including housing and infrastructure, remained inadequate.

The scholar observed a recognizable inequality in Rio's urban planning. Additionally, Pereira et al. (2017) reflected on the institutional challenge, highlighting a weakness in the interoperability of data sources across several agencies, which created serious delays and blind spots.

Local departments continued using their separate systems, and the Operations Center did not fully apply the lessons learned in its daily operations. In addition, an imbalance in cost and benefit was identified. The operating expenses of the center were high, while the city failed to address systemic problems, such as poor drainage or unstable hillside housing. International observers noticed that massive investments in high-tech control rooms were not balanced with infrastructure upgrades that could directly benefit residents. The lesson here is that Smart City perspectives must be paired with equitable policies and systematic reforms, strongly emphasizing inclusive urban progress, rather than merely serving as a symbolic showcase.

Many developing nations have financed ambitious smart city projects, hoping to leap into a high-tech future. Yet, because these technologies were deployed without local buy-in or attention to real conditions, underperforming projects like Jakarta's Smart Grid Pilot and Rio de Janeiro's Operations Center demonstrate that ignoring context can undermine well-funded initiatives.

A shared lesson from the two projects is that the deployment of advanced tools without addressing local capacity, maintenance, and equity can produce poor outcomes. A smart grid is useless if it is not fully trusted or understood by city residents. A command center can achieve little if it is not designed to serve the broader community. Smart City initiatives must start with community engagement, clear maintenance plans, and integration with essential urban services.

2.4 Redefining Smart: Beyond Towers and Tech

For many years, when we discuss "Smart Cities," we often visualize images of shining skyscrapers, central command units, and costly digital platforms. However, in many developing countries, these images have proven overly ambitious, because city leaders focused on digital solutions rather than on what truly matters most.

To build a real Smart City for everyone, it's important to redefine what "smart" means. The future of Smart Cities belongs not only to gadgets but also to those who deploy technology intelligently, formulate policies supported by data, and engage communities as active partners.

Figure 2.2: Smart City – Beyond Towers and Tech

i. The Shift to Contextual Smartness

We cannot call an urbanized city "smart" simply because it installs sensors everywhere, applies a 5G network, or constructs symbolic towers. True smartness can be understood as urban management that responds effectively to local conditions. The idea of "contextual smartness" is a novel approach.

It is not about replicating Singapore's skyline or Songdo's command centers; it is about ensuring that every investment delivers real benefits through the alignment of tools, policies, and community needs. In many low- and middle-income nations with tight budgets and uneven infrastructure, contextual smartness prioritizes efficiency and practicality over showpieces.

Instead of spending heavily on a citywide autonomous transit system, a municipality should start by using basic GPS data to redesign bus routes and reduce waiting times. Rather than inaugurating a costly new control center, mayors should focus on interconnecting existing data sources and creating a shared dashboard to allocate resources for waste collection and street maintenance.

ii. Efficiency as a Smart Value

Efficiency should be fully understood as saving budget and improving citizens' quality of life, not just glamorous design. Governments might boast of a smart city future with drones flying around impressive iconic towers, but more practically, districts should upgrade drainage systems, utilize rainfall data to prevent floods, equip adequate lighting systems, and provide essential public services.

A city is really "smart" when it can optimize funds, spend wisely on the basic needs of the locality, and not just boast about a grand plan. A mid-size city in Southeast Asia should consider using a low-cost sensor to track water leaks through open-source mapping. As a result, after one year, it could reduce wasted water by 20%, demonstrating the significant impact of smart, context-driven strategies.

iii. Data-driven Governance

A truly Smart City is not defined by the number of technologies it deploys; it should be defined by how it analyzes and applies information to respond promptly. Data-driven governance means the city is managed according to a culture of data-driven rules and evidence-based decisions, without political interference. This discipline is particularly critical in many developing countries due to their resource limitations.

In low- and middle-income nations, cities collect a great deal of information on traffic records, flood reports, health statistics, and even budget rates, but sometimes this data is not properly used and remains fragmented. Contextual smartness urges the transformation of information into realistic actions for residents' daily lives. For example, accurate health statistics can mobilize funding for clinics and vaccinations, real-time traffic data can help municipal managers alert and redirect transportation routes before a typhoon, and local portals can show residents how funds are allocated and spent, thereby enhancing transparency. Otherwise, communities can be passively left behind. A powerful approach does not necessarily require the costliest or most state-of-the-art technologies; conversely, basic tools, when combined with genuine commitment and intelligent data, can enhance governance. The principle is to collect information and integrate it into daily decisions. City departments should develop teams capable of analyzing, predicting, and sharing data. Then, they should communicate the results and be responsive to residents. We affirm that technology is a tool employed by smart people who manage a truly Smart City.

iv. Community Engagement as a Core

The most fundamental foundation is community engagement in the co-design process for the city's future. Smart City projects often follow a top-down approach, without sufficient consultation or evaluation of public opinions. Instead of imposing decisions, contextual smartness aims to accurately assess local needs and gather feedback on solutions by involving the community in all steps of project design. Let's review the case studies in Latin America, where numerous municipalities encourage their citizens to use participatory budgeting apps. Porto Alegre, Brazil, pioneered participatory budgeting during the planning stage, with voting procedures conducted through digital platforms.

In Colombia, the cities of Medellín and Bogotá have enabled citizens to exercise their right to participate in political decision-making through dedicated digital platforms. Similarly, in Peru, Lima and Cusco have experimented with online participatory budgeting initiatives that allow residents to vote on small-scale infrastructure projects, such as drainage improvements and public lighting. In Lima, city administrators introduced the consulta virtual, an online consultation and voting mechanism integrated into the participatory budgeting process. These digital tools not only broaden civic participation but also foster a sense of community ownership, enhance citizens' understanding of their rights, and strengthen trust between residents and municipal leadership.

vi. Broader Definition - Inclusion

To redefine "smart," we mean using technology to serve people and community needs. A city with limited funds can still spend strategically, effectively, and wisely, applying solutions in the appropriate context. Cities should not import vastly unfamiliar technologies or rely excessively on external consultants; instead, they should listen to residents and adapt to real conditions. Pilot outcomes should be carefully evaluated using statistical methods, focusing on metrics such as reduced energy losses, shortened commutes, lowered crime rates, or cleaner streets. These elements are more important than flashy tech buildings, because developing nations need tangible, practical improvements.

v. Recommendations to Be Done

For contextual smartness to be effectively embraced, developing countries should take the following immediate actions:

- **Audit and Optimize:** Assess the city's current data and existing infrastructure that may be ready for usage.
- **Engage and Educate:** Involve the community in the project plan, raise public awareness, and train residents/local technicians in using tools.
- **Scale strategically:** Conduct pilots on project components, measure solutions, and roll out if they are proven trustworthy and sustainable.

2.5 Hidden Strengths: What Developing Cities Already Bring to the Table

When developing countries discuss creating Smart Cities, controversial issues are often raised about what is lacking, such as funds, infrastructure, or modern technology. However, what is often overlooked are existing valuable assets, which can enable smart development at lower cost. The first step should be recognizing internal strengths and building confidence; the grand vision for a megacity may need to be approached more cautiously to ensure successful urban progress. Firstly, consider the strength of community networks. Networks are powerful channels for collecting data, reporting, and planning public services. For example, in cities in Sub-Saharan Africa and Southeast Asia, solidarity and neighborhood connections support the creation of such networks. Secondly, consider whether cities have a significant youth population. Demographic analytics relate to tech-savvy individuals and their openness to adopting advanced technology. For example, in Lagos, Ho Chi Minh City, and Manila, young people are ready to embrace digital platforms and innovations. Thirdly, consider the rate of mobile device usage. A high rate of smartphone usage enables the deployment of low-cost apps for reporting issues, mobile banking, and more responsive services compared to wired networks.

Figure 2.3: Considering how to implement Smart City Strategies

Fourthly, consider consumer behaviors and cultural adaptability. Informal street vendors, transit operators, local knowledge, and flexibility should be incorporated into smart initiatives, which should include clear and practical solutions. Southeast Asian countries often experience a rapid pace of urbanization, transitioning from traditional practices to a hybrid model. In Sub-Saharan Africa, adaptation is often informal as communities settle and organize gradually. No single formula applies to all regions; each region has its own strengths that can support a Smart City. Finally, consider existing assets and resources. Instead of copying the high-profile models from high-income countries, Smart City projects should be designed around a tailor-made network, community engagement, and affordable, inclusive solutions that fit the local context.

i. An in-depth study about a relevant case: Vietnam's policy-making process for Smart Cities and recommendations

To keep up with the trends of the Smart City era, Vietnam has identified Smart City development as a key national growth strategy. Vietnam is progressing in smart urban planning, in alignment with the implementation of Goal 11 of the 2030 Agenda for Sustainable Development, which aims to create inclusive cities: "developing sustainable and resilient urban and rural areas; ensuring safe living and working environments; and proportionally distributing population and labor by region". Concurrently, Vietnam has taken strong actions in the institutional and policy-making process by issuing relevant documents to direct and guide the essential criteria for Smart City construction and development:

- Decision 1819/QD-TTg, dated May 26, 2015, of the Prime Minister approving the National Program for the Application of Information Technology in State Agencies' Activities from 2016 to 2020, which mandates "deploying Smart Cities in at least three locations according to the criteria guided by the Ministry of Information and Communications".
- Official Dispatch No. 10384/VPCP-KGVX, dated December 1, 2016, by the Government Office, regarding the development of sustainable Smart Cities globally and in Vietnam.

- Document No. 58/BTTTT-KHCN, dated January 11, 2018, by the Ministry of Information and Communications, providing guiding principles for information and communication technology in Smart City development in Vietnam.
- Decision 950/QD-TTg, dated August 11, 2018, by the Prime Minister, approving the Project on Sustainable Smart City Development in Vietnam from 2018 to 2025, with orientation toward 2030 (Project 950).
- Decision 829/QD-BTTTT, dated May 31, 2019, by the Ministry of Information and Communications, promulgating the ICT reference framework for Smart City development (version 1.0).
- Document No. 3098/BTTTT-KHCN, dated September 13, 2019, from the Ministry of Information and Communications, announcing the Vietnam Smart City Index to 2025 (version 1.0).
- Resolution No. 52-NQ/TW, dated September 27, 2019, by the Politburo, on policies and strategies to proactively participate in the Fourth Industrial Revolution, setting the goal of forming smart urban chains in key economic regions of the North, South, and Central regions, gradually connecting with regional and global smart urban networks.
- Document 693/BXD-PTDT, dated February 21, 2020, by the Ministry of Construction, on the implementation of sustainable Smart City development in Vietnam.
- Decision No. 06/QD-TTg dated January 6, 2022, by the Prime Minister, approving the project for developing applications of population data, identification, and electronic authentication to serve national digital transformation in the period 2022-2025, with a vision to 2030.
- Resolution No. 06-NQ/TW, dated January 24, 2022, by the Politburo, on the planning, construction, management, and

sustainable development of Vietnamese urban areas through 2030, with a vision toward 2045.

- Decision No. 505/QD-TTg, dated April 22, 2022, establishing National Digital Transformation Day.
- Document 2333/BTTTT-CĐSQG, dated June 20, 2023, by the Ministry of Information and Communications, on the implementation of ICT for Smart City development and the establishment of Smart Monitoring and Operation Centers (IOCs) in localities.
- Resolution 57/NQ-TW dated December 22, 2024, by the Politburo on breakthroughs in the development of science, technology, innovation, and national digital transformation, with the goal of completing the construction of Smart Cities by 2030 for centrally-administered cities and some provinces and cities that meet the conditions.
- Decree 269/2025/ND-CP dated October 14, 2025 by the Government on the development of Smart Cities.
- Decision No. 2597/QD-TTg dated November 27, 2025, by the Prime Minister approving the digital transformation plan in the construction, management, and exploitation of construction infrastructure.
- Regional and local documents on Smart City construction in Vietnam (Binh, 2023).

To comply with the legal framework and policies, Smart Cities in Vietnam are advised to implement the following recommendations:

➢ When developing Smart City projects, Vietnam should adopt a people-centric approach, focus on cultural factors, follow modern trends, and respect tangible and intangible heritage as well as natural values. From the project planning stage, Vietnam needs to strengthen communication and raise awareness among businesses, social organizations, and residents regarding their responsibilities, opportunities, and roles in building Smart Cities (Diem, 2021).

- Vietnam should pay attention to connecting regional transportation routes, saving energy, minimizing and managing waste, preventing crime, ensuring traffic safety, and improving people's work efficiency through the use of Big Data and the Internet of Things.
- Vietnam should create a flexible and robust development mechanism for the smart housing market and smart real estate transaction tools. In the current Vietnamese market, smart urban real estate products can be considered a niche market, where buyers' preferences are increasingly diverse, requiring product diversification to cater to different housing segments and customer demands, thereby benefiting a wider range of people. In the new trend of housing concepts, there is a shift from viewing housing merely as accommodation to seeing it as a symbol of social status and personalization. Smart home models also need to distinguish themselves from traditional housing. Therefore, smart home business models must differ from conventional business models. A popular smart business trend is the use of digital business platforms in urban development. Vietnam should issue policies and mechanisms to encourage the development of smart grid solutions for Smart Cities, such as rooftop solar power systems, energy storage systems, and modernized distribution grids.
- Vietnam needs to train and develop high-quality human resources to support the urban digital transformation process, build capacity for Smart City development according to appropriate standards, foster smart citizen communities that live and work efficiently, and encourage research and innovation through practical inventions and patents. Expert groups need to strengthen their activities,

develop and evaluate technical standards for Smart Cities, and provide specific policy recommendations in a timely, continuous, and incisive manner.

- Smart urban policy planning needs to be consistent, promote administrative reform, and balance the development of Smart Cities with social inclusiveness, ensuring that people are guaranteed equal rights rather than merely emphasizing technology. Concrete strategies should include community engagement beyond apps and the implementation of affordable sensor technologies. In the trend of smart data management, information security and safety are critical. Therefore, Vietnam should promptly introduce robust policies and institutional measures to protect citizens' privacy while ensuring strong security systems. Vietnam should prevent the illegal collection of personal data, the abuse of information, and the misuse of data. The government and relevant sectors need to regulate and ensure user safety and convenience through specific measures and policies. These measures will strengthen the legal environment and build trust in smart, connected technologies. Security and privacy measures should be fully integrated from the very beginning of the development process for new smart home technologies.
- The Government should promote and invest in research, as well as apply advanced science and technology in Smart City development; encourage innovation and the creation of Smart Cities based on new technology platforms, such as Big Data, AI, blockchain, IoT, cybersecurity, clean energy, and environmental technologies. Furthermore, the Government should support the building of networks to exchange experiences, share knowledge, and promote Smart Cities that advance sustainability, environmental protection, and climate adaptation.

- Smart from each mega project to building or house/apartment.

2.6 Smart on a Budget: Leveraging What Works

Smart Cities are often envisioned with heavy investments in control centers, sensor installation, and smart homes. These visions seem to be beyond the reach of many developing nations. However, with limited finances, cities can certainly approach budgeting differently, concentrating on efficiency, creativity, and the smart use of available resources. The motto "smart on a budget" emphasizes operating at the right scale, implementing frugal innovations, and achieving visible results.

i. Using What Already Exists

The city of Manila, Philippines, partnered with telecom providers to create a text-based early warning system for floods and traffic congestion. Manila utilized existing cell towers, public reports, and free social media, rather than investing in high-cost infrastructure.

During storms, the timely warning system alerted districts in time to help residents evacuate and reroute traffic. These measures saved lives and reduced danger for thousands of people (Porio, 2011). Another example is Nairobi, Kenya. Nairobi's matatus, informal minibuses, are the most common form of local transportation.

Matatus shares GPS data through low-cost mobile apps, integrated with mobile ticketing machines and cameras. Kenyans can track buses and plan trips through a real-time system, a significant improvement given that 70% of Nairobi's population depends on matatu buses. This solution is adaptive and does not require large funds, as it utilizes existing transport operators and simple technology for smarter commuting (CNN, 2025).

ii. Low-Tech Infrastructure Wins

A Smart City does not always need the most cutting-edge solutions, and admittedly, machines cannot replace people. In many developing countries, some of the most effective improvements are derived from low-tech infrastructure that is planned thoughtfully and that involves the community as co-designers. These initiatives should be more creative and context-specific, rather than depend on costly hardware.

In the area of waste management, developers can recognize the existing network of informal waste pickers instead of importing high-cost waste-sorting machines. The city is encouraged to train cooperative teams, establish a system for monitoring recycling rates, and simultaneously improve people's livelihoods. The city can also establish collection points and well-coordinated routes. Although this solution is low-tech, the results can outperform many high-tech systems in other developed countries. Regarding the flood solution, city districts can train community volunteers to monitor the water levels of lakes or rivers via reporting apps on smartphones instead of installing thousands of expensive sensors.

Additionally, the timely interpretation of data within the municipal dashboard can generate warnings for dangerously high-water levels. Through such a culture of responsibility fostered by an inexpensive system, significant flood damage can be reduced. What should a transport planner do? They can plan affordable cable cars to connect directly to the metro instead of waiting for budgets of millions of dollars for a subway or highway. This method may transform residents' lives, reduce commute times, and open new job opportunities.

It can be argued that low-tech solutions do not necessarily mean low impact in many cases. This approach involves brainstorming for the city's needs, identifying areas for improvement, and ensuring systematic long-term maintenance, rather than focusing on massive investments.

Local workers, volunteers, and residents will feel that they are truly the owners of a well-functioning system that they should care for and adapt to, despite the hardship. In this "smart" era, context-appropriate infrastructure is the most effective investment for the city.

iii. A Free Price Scheme of Digital Participation vs. Frugal Innovation

The participation of the community in the city's decisions does not depend on high-tech solutions. It can be facilitated through online voting or mobile apps. Digital platforms require only smartphones and Wi-Fi, making them accessible and affordable for all. Citizens can even report service issues through WhatsApp, and the feedback workflows go directly to the mayors without heavy investment in platforms.

Frugal innovation can be understood as a city's ability to solve problems using available resources. The purchase of high-cost systems from the most advanced countries must take into account whether they are realistic and applicable in the local context. Cities can utilize existing networks, informal personnel, low-cost software, and hardware, all aimed at achieving efficient solutions. Projects should first be tested for proven effectiveness and then for local applicability.

2.7 Global alarming issue and Advice to the developing countries when adopting Smart City strategies

i. Major emerging issues

In the latest report, IMD Smart City Index 2025 impresses the global reader community with the theme *"The housing affordability challenge: a growing concern"*. The report flags housing affordability as the central and growing concern for 2025, not only for a single nation or city, but also spans across countries.

Figure 2.4: Affording a house is increasingly difficult

A survey is conducted with 146 cities, among those, respondents from 110 cities, which have at least half of the respondents in each city, answer that affordable housing should be the top priority. Over the past decades, housing has become increasingly unaffordable for many households. This trend is reflected in the IMD Smart City Survey: in Lisbon, Vancouver, and Dublin, some nine out of ten respondents report having difficulties finding housing that costs less than a third of their income (IMD, 2025).

ii. Advice to developing countries when adopting Smart City strategies

Similar to developed countries with high-profile Smart Cities, such as Vancouver, Dublin, or Dubai, making housing more affordable is the most urgent priority for developing countries as they pursue mega Smart City projects. Housing strategy should be embedded in Smart City strategies and should not treat technology as the only core pillar. A smart plan cannot ignore the housing affordability target.

Governments should align national policy frameworks (subsidies, fiscal transfers, land policy) with city-level smart plans and design targeted subsidy programs for lower-income groups, such as tailored finance products, rental subsidies, first-time homebuyer or social housing schemes, aimed at preventing households from financial shocks.

Government policies should be designed with high flexibility toward the real estate market, avoiding extreme measures like those of the Chinese government. In August 2020, China took steps to curb excessive leverage in real estate and de-risk the sector. Specifically, the "Three Red Lines" policy in real estate corporate finance required companies to achieve lower debt ratios to limit reliance on borrowed capital and minimize bubble risks by controlling property prices.

As a consequence, leading developers like Evergrande and Country Garden faced "behind schedule" and "liquidity shortfall" projects, profoundly impacting the Chinese real estate sector (Yang, 2023). According to IMD (2025), developing countries should be prudent when adopting "smart policies" for Smart Cities, focusing on attracting high-skilled immigrants to become talent-importing cities. The surge of international professionals contributes to a city's innovation but also leads to a sharp rise in rental costs and housing prices amid constrained supply. In shaping such smart policies, valuable insights can be drawn from the experiences of cities such as Zurich, Geneva, and Dubai. In Dubai, more than 40% of residential property value is held by foreign nationals, a result of a 20% increase in foreign-owned residential real estate during a short period from 2020 to 2022. Therefore, affordable housing becomes a greater challenge for local low-income and middle-class households. An open immigration policy will be more effective when accompanied by measures ensuring equity.

In shaping the urban future from "Connected Cities" to "Cognitive Cities" smart city initiatives in developing countries must move beyond passive connectivity based solely on the Internet of Things (IoT). Instead, they should adopt artificial intelligence as a true urban operating system, capable of self-learning, self-adaptation, and autonomous decision-making. In this cognitive city model, data and AI are used proactively to anticipate and meet residents' needs rather than merely reacting to events, with the shared ambition of effectively leveraging up to 90 percent of available data.

iii. Solving the global concern: housing affordability

It is noted that construction cost is the main factor influencing the sales price of real estate. The demand for cement, steel, glass, and skilled workers is affected by inflation, especially when there are shortages in materials, labor, or disruptions in the supply chain. In 2025, the National Association of Home Builders (NAHB) conducted a nationwide survey of approximately 4,000 home builders across the United States.

According to this survey, in 2024, the average sales price for a new single-family home in the U.S. was $665,298, with construction costs alone accounting for 64.4% of the sales price, and 13.7% attributed to finished lot costs (NAHB, 2025). This implies that developing nations may face a double burden, including high import costs for conventional materials and limited domestic production capacity.

From the authors' perspective, controlling construction costs for Smart Cities' housing is critical; however, all city builders must understand and rethink how houses should be constructed. When building Smart Cities, developing countries are urged to find alternatives to traditional materials, construction methods, and current labor resources, as suggested in the following.

2.7.1 3D-Printed House: An Innovation in Construction

The application of 3D printing technology in building houses can generate low-cost housing rapidly, using less labor and producing less material waste, making it a promising approach for sustainable construction. 3D-printed houses are a breakthrough for developed countries such as the United States, the Netherlands, Germany, Denmark, and the UAE. A 3D-printed house is not the "House of the Future"; it is the "House of the Present." Various developing countries have succeeded in 3D-printed housing projects, such as Mexico, India, Indonesia, Kenya, and Vietnam.

Figure 2.5: 3D-Printed House

While developed countries lead in design, material R&D, and regulation, developing countries adopt 3D printing technology as a low-cost and fast solution for housing affordability, particularly in the context of sharply rising material prices and labor shortages. The Mexican company MANUFACTURA uses 3D printing technology to create building materials from discarded eggshells.

The "Eggshell" Project addresses two problems: the negative environmental impact of the ceramics industry and the waste of food and other organic materials in Mexico. The ceramics industry generates large emissions due to the use of fossil fuels, and the waste from the production process is also significant. Therefore, MANUFACTURA decided to reuse lightweight, readily available eggshells from waste for construction technology. When combined with a bio-based adhesive, eggshells create a mixture that can solidify without firing, making it ideal for 3D printing.

2.7.2 Prefabricated and Modular Construction

In Singapore, Sweden, and Japan, the modular housing system is a major focus of the National Housing Strategies. For developing nations, modular construction can minimize on-site labor, which is especially important when skilled workers are scarce and highly paid. Prefabricated components manufactured off-site reduce waste, save time, and lower labor costs by streamlining assembly.

Prefabrication can standardize mass-produced housing units that meet safety and energy requirements, helping governments deliver affordable homes more efficiently, durably, and sustainably. Prefabricated and modular construction is fully scalable in developing countries.

2.7.3 Engineered Bamboo and Timber

Engineered bamboo and timber represent a low-carbon, aesthetically appealing trend for building affordable and resilient houses in Smart Cities. Bamboo and cross-laminated timber are considered ideal alternatives to steel and concrete, as they are renewable, lightweight, and carbon-friendly.

While developing nations struggle with high material costs and limited local resources, engineered bamboo is low-cost and eco-friendly; after being treated and laminated, it becomes stronger than many hardwoods and can rival steel in tensile strength.

This makes it an excellent solution for tropical and subtropical regions of Asia, Africa, and Latin America. Countries like Vietnam and Indonesia are leading in bamboo-based prefabricated housing systems that reduce carbon emissions, local construction costs, and embodied energy.

2.7.4 Recycled and Low-Carbon Materials

In Smart City development, adopting recycled and low-carbon materials such as recycled aggregates, fly ash concrete, and geopolymer cement can reduce dependence on traditional cement and sand, whose extraction often strains the environment. These substitutes target emission cuts by up to 40% and lower overall material costs when locally sourced.

This solution is regarded as more affordable, sustainable, and adaptable than others. Cities are built smarter, greener, and more inclusive when combined with digital design tools and prefabrication techniques. These materials support rapid urban housing growth strategies with reduced environmental damage. Currently, India, Kenya, and the Philippines are experimenting with recycled plastic-based housing solutions that are not only low-cost but also highly resilient to weather and climate stresses.

These innovations demonstrate how waste-to-housing technologies can make Smart City development more inclusive and sustainable, especially in regions facing rapid urbanization and resource scarcity.

2.7.5 Local Material Optimization

Local material optimization refers to the strategic use of readily available, region-specific building materials to reduce construction costs, environmental impact, and supply chain dependency. Instead of relying on imported or high-tech materials, this approach focuses on maximizing the potential of what is locally abundant, whether clay, laterite, stone, bamboo, or agricultural by-products.

Promoting the use of local materials is essential for reducing logistics costs and supporting local employment opportunities. These methods, combined with digital design tools, can ensure modern aesthetics while maintaining affordability. In Vietnam, the

"Container House" is highly mobile, low-cost, and easy to install, making it particularly suitable for cool-climate regions.

2.7.6 Digital Construction Management and BIM (Building Information Modeling)

These represent transformative tools in modern urban development. Urban experts use the above-mentioned models to help cities plan, design, and deliver housing and infrastructure more efficiently. These technologies bridge the gap between traditional construction practices and the data-driven demands of Smart Cities. Using BIM systems improves coordination, reduces design errors, and optimizes material usage.

By preventing delays and resource waste, BIM indirectly reduces total project costs and creates an intelligent 3D model for a building, integrating detailed information about its materials, structure, costs, and performance throughout its lifecycle, including energy efficiency and performance in various climate conditions.

2.7.7 Labor Upskilling and Technological Training

Labor upskilling and technological training form the foundation of human resources for Smart City development. Without a skilled workforce, even the most advanced technologies cannot deliver sustainable, affordable, and inclusive urban growth. To address labor shortages, governments should prioritize investment in vocational training and digital construction tools.

Partnerships between governments, universities, and the private sector can ensure a skilled workforce ready for modern, efficient building techniques. Prefabrication, 3D printing, and digital management systems require extensive teams of engineers, architects, and policymakers who can maximize a city's benefits through higher construction quality, faster project completion, lower maintenance costs, and enhanced local employment.

2.8 Conclusions

For developing nations, it can be argued that a city does not need to mirror New York's skyline or Amsterdam's sensor networks to be smart. Instead, cities need to use financial resources wisely, listen to residents' opinions, gather feedback, and design projects that fit the local context, mitigating risks by utilizing available resources.

With the motto "smart budgeting," cities can deliver cleaner streets, more sustainable transportation, quicker emergency responses, and more inclusive decision-making, rather than simply waiting for funding approval. Furthermore, urban planners should not view Smart Cities as a fantasy or a marketing tool used by real estate developers to sell high-end houses and apartments at extremely high prices.

Cutting-edge technologies should not be used to position the housing market in a luxury segment serving only the wealthy. If smart city designers maintain that mindset, they have overlooked fundamental principles of economic management. Disruptions in the real estate supply chain can lead to a market crisis and increase geopolitical tensions if even a single economic factor falls out of balance.

We have witnessed the fragile market resulting from giant Smart City projects in many developing countries across Asia, Africa, and Latin America. Countless houses and buildings have been abandoned due to governments' neglect of local demand, while oversupply challenges both geopolitical and societal stability.

The authors call for inclusive implementation, recommending that Smart Cities should be planned with the community at the center and provide comprehensive digital solutions to address pressing social issues, such as urbanization, aging, unemployment, traffic congestion, environmental protection, and security. Smart Cities, in a more humane approach, should focus not only on a race for technological advancement but also on developing a comprehensive ecosystem that creates livelihoods for grassroots communities and job opportunities for regional sustainable development. Most importantly, housing affordability should be the top priority for developing countries when shaping a Smart City agenda.

A city is truly smart when its residents can afford basic shelter, and social resilience should be built not only on digital transformation but also on housing affordability. Integrating alternative materials, industrialized construction, and digital project management can significantly reduce costs while supporting sustainable growth and local resilience.

References

Binh, L. T. (2023). Perspectives in developing sustainable Smart Cities in the new context. *Developing Smart Cities in the world and lesson learnt for Vietnam* (p. 15). Hanoi: Finance Publisher.

CNN. (2025, July 25). Nairobi's matatu minibuses are getting a high-tech makeover. Retrieved from https://edition.cnn.com/2020/01/28/business/data-integrated-kenya-intl

Diem, T. N. (2021). Developing sustainable Smart Cities: Study on cities in Japan. *Journal of North East Asia*, 51.

EPA. (2025, July 24). U.S.-Brazil Joint Initiative on Urban Sustainability.https://archive.epa.gov/international/jius/web/html/rio_operations_center.html

IMD. (2025). *World Competitiveness Center*. https://www.imd.org/smart-cityobservatory/home/rankings/

Le, V. T. (2023). Developing smart cities in China and recommendations for Vietnam. *Developing Smart Cities in the world and lesson learnt for Vietnam* (p. 23). Hanoi: Finance Publisher.

Loso Judijanto, S. (2024, December). The Effect of Smart Grid Implementation on the Integration of New Renewable Energy in Indonesia's Electricity System. *West Science Nature & Technology, 2*.

Moutinho, N. e. (2016). The Rio Operations Center: Promise and Limits. *Analyzes how the center prioritized central districts over informal areas., 35(3)*. IEEE Technology and Society Magazine.

NAHB. (2025). *Cost of Constructing a Home-2024* . Washington D.C: National Association of Home Builders

Nduru, J. (2021, July). *Kenya's Konza Techno-City flawed and floundering Neoliberal Urbanism*. https://www.researchgate.net/publication/353104155_Kenya's_Konza_TechnoCity_flawed_and_floundering_Neoliberal_Urbanism

Newswire, P. (2019, July 30). *Forest City "Smart B.I.A"*

System Wins IDC Smart City Asia Pacific Awards. Retrieved from https://www.prnewswire.com/in/news-releases/forest-city-smart-b-i-a-system-wins-idc-smart-city-asia-pacific-awards-892494967.html

Pereira, G. E. (n.d.). Delivering Smart City Strategies in Brazil: A Maturity Model Based on Institutional Indicators. *Discusses Rio's operations center and uneven benefits*. Information Polity. doi:10.3233/IP-170039

Porio, E. (2011). Climate Change Vulnerability and Adaptation in Metro Manila. *Asian Journal of Social Science, 39 (4)*, 425–445. doi:https://doi.org/10.1163/156853111X597240

UNHABITAT. (2025, July 22). https://unhabitat.org: https://unhabitat.org/sites/default/files/documents/2019-05/rwanda_smart_city-master_plan.pdf

UN-Habitat. (2025, July 22). *UN-Habitat, Rwanda Smart City Masterplan (2017).* Retrieved from p. 10–12, p. 22–23, p. 34–35: https://unhabitat.org/sites/default/files/documents/2019-05/rwanda_smart_city-master_plan.pdf

WorldBank. (2024, Aug 1). Middle-Income Trap" Hinders Progress in 108 Developing Countries. https://www.worldbank.org/en/news/press-release/2024/07/22/-middle-income-trap-hinders-progress-in-108-developing-countries

WorldBank. (2025, July 22). The World Bank In Malaysia. https://www.worldbank.org/en/country/malaysia/overview

Yang, e. a. (2023). The Impact of Three Red Lines Policy on Real Estate Industry. *The 2nd International Conference on Business and Policy Studies.*

CHAPTER 3
The Smart City on a Budget: Affordable, Practical Solutions

3.1 Introduction

Africa is projected to be home to over 2.5 billion people by 2050, with nearly 55% expected to migrate to urban areas (Güneralp et al., 2017). This significant urban influx will place immense pressure on existing infrastructure. This situation highlights the urgent need for leaders and stakeholders to collaborate on synergistic approaches that can drive infrastructure development to support this migration (OECD/SWAC, 2020). Given that infrastructure development requires substantial capital investment, it is essential for governments to partner with stakeholders to pool resources and expand infrastructure capacity to accommodate the surge in urban migration.

The role of the central government is crucial in providing leadership, policy direction, and strategic vision for successful smart city development (Uyarra et al., 2020). Ultimately, a "smart city on a budget," when translated into practical and affordable solutions, can significantly enhance urban living. Take Amina, for instance, a savvy resident of Kibera, one of Nairobi's largest informal settlements.

Her day doesn't start with a high-tech home assistant; instead, she begins by quickly checking her mobile phone. Using a straightforward app, she provides feedback about overflowing waste bins directly to local authorities, prompting a community-run waste management team to respond. By evening, her street is clean, without costly infrastructure or advanced technology deployment. This innovative method of waste management is crucial because it demonstrates that environmental sustainability can be integrated into urban living without massive infrastructural development. One of the core principles of a Smart City is environmental sustainability, which is exemplified by the initiatives of Kibera's residents. The Kibera strategy is well-suited for Smart City initiatives due to its simplicity, low cost, and scalability, and it can be replicated by other developing countries.

Citizen engagement plays a vital role in smart city initiatives because it effectively identifies community needs in a more accessible way. As urbanization continues to grow globally, leaders are increasingly adopting innovative strategies to transform their communities into smart cities that can respond effectively to this surge. This transformation involves reengineering infrastructure to facilitate a smooth transition.

However, moving toward a Smart City can impose significant costs on central governments, making it challenging to be managed solely with public resources. Smart City initiatives demand considerable resources; therefore, partnering with the private sector is essential to distribute the financial burden and share risks (Clark, 2020). This is the essence of a "Smart City on a budget," which emphasizes practical, affordable solutions that can lead to a much better quality of urban life.

3.2 Low-Tech, High-Impact: Tools That Work Anywhere

Smart Cities are often envisioned as the culmination of advanced technologies and substantial investments in infrastructure. They typically feature extensive Internet of Things (IoT) systems, complex sensor networks, and large data-management centers. While these elements are indeed transformative, they can be prohibitively expensive and beyond the reach of many developing countries (McKinsey Global Institute, 2018). Consequently, fostering smartness within a resource-constrained environment necessitates the use of accessible, cost-effective, yet high-impact technologies. Low-tech, high-impact tools present viable alternatives that enable cities to transition toward smart city frameworks. These tools can serve as catalysts for the Smart City concept without the need for massive infrastructure development or significant budget allocations from central governments.

These tools can focus on leveraging existing resources and engaging communities to enhance urban living. Here, we highlight practical and affordable technological solutions that have proven effective in improving urban living under constrained budgets. As a result, smartness in a resource-constrained environment must be driven by available, cost-effective, yet high-impact technologies. Here, we outline pragmatic and inexpensive technological solutions that have demonstrated significant urban benefits, even with severe budget limitations.

i. Mobile Applications for Citizen Engagement

The growth of mobile phones, especially smartphones, in developing countries offers a remarkable opportunity to enhance urban governance at a minimal cost. Citizen Engagement Mobile Applications (CEMA) serve as vital platforms that allow residents to report urban issues directly to the appropriate authorities, fostering transparency, accountability, and responsiveness (GIZ, 2018).

These models of citizen engagement are essential for driving Smart City development and encouraging community ownership. Citizen involvement helps identify needs and concerns, promotes accountability and transparency, reduces the potential for corruption, and ensures that resources are used effectively. Engaging citizens not only empowers them but also builds stronger and more resilient urban communities.

ii. Case Study: Ushahidi in Nairobi, Kenya

Originally created to document violence during Kenya's disputed 2007 election, Ushahidi has transformed into a powerful tool for enhancing urban governance. The platform allows citizens to report real-time issues such as potholes, waste accumulation, crime, and infrastructure damage via SMS or a mobile app (Okolloh, 2009).

In Nairobi, local authorities have effectively utilized Ushahidi to quickly identify and address municipal problems, significantly boosting responsiveness and efficiency. The Ushahidi implementation has reduced response times from days or weeks to just a few hours, fostering greater public trust and enhancing the city's service delivery without substantial financial investment in infrastructure (GIZ, 2018).

The term "Ushahidi" itself means "testimony" or "witness" in Swahili, reflecting the platform's core mission as a crowdsourcing tool. Ushahidi was created in 2008 to enable real-time information sharing and problem resolution through Short Message Service (SMS). The impact analysis of the Ushahidi project is shown in the table below:

Table 3.1: Impact Analysis of the Ushahidi project
Source: GIZ (2018)

Metric	Before Ushahidi (2008)	After Ushahidi Implementation (2012)
Municipal Response Time (Average)	14 days	2 days
Citizen Engagement (Reports/month)	50	600
Citizen Satisfaction	20%	78%

iii. Summary and Relevance

Precise urban mapping and reliable data are essential for effective urban planning and disaster response management. However, in many developing countries, detailed geospatial data is limited and costly. The Ushahidi platform represents a distinctive model of a Smart City initiative in Africa, specifically designed to utilize scarce resources efficiently. This approach demonstrates how

Ushahidi residents and authorities harnessed simple technology to foster citizen-driven innovation. Importantly, a Smart City leverages technology and data to improve areas such as citizen engagement, urban service delivery, waste management, and traffic control. These models align closely with the goals of the Ushahidi platform. Ultimately, Ushahidi stands out as a cost-effective model that maximizes local knowledge and resources to achieve tangible results.

3.3 Open-Source Mapping and Crowdsourced Data Collection

Precise urban mapping and reliable data are essential for effective urban planning and disaster response management. Unfortunately, detailed geospatial data is often limited and costly in many developing countries.

Open-source platforms like OpenStreetMap (OSM) help bridge this gap by allowing community members to actively contribute to the creation of accurate maps, enhancing accessibility and precision in urban planning efforts. This collaborative approach empowers local communities and improves overall data availability.

i. Case Study: Dar Ramani Huria Project, Dar es Salaam, Tanzania

The rapid, unplanned growth of the city, coupled with inadequate drainage systems, significantly increased Dar es Salaam's vulnerability to seasonal flooding. The Dar Ramani Huria project recruited local university students and residents to help map areas susceptible to flooding (Iliffe et al., 2017).

The maps produced through this initiative not only assisted urban planners in recognizing vulnerabilities but also facilitated the implementation of targeted, low-cost interventions that dramatically reduced flood risks. The impact analysis of the Dar Ramani Huria Project is captured in Table 3.2.

Table 3.2: Impact Analysis of Positive Urban Mapping projects in Dar Ramani Huria, Source: Iliffe et al. (2017)

Metric	Before Mapping (2014)	After Mapping Implementation (2018)
Flood response effectiveness (%)	40%	85%
Affected households	~50,000/year	~15,000/year
Cost savings (annual)	N/A	$800,000

ii. Summary and Relevance

The Dar Ramani Huria initiative in Dar es Salaam approach to mitigate flooding aligns closely with the principles of a smart city initiative. Through this initiative, community members and various stakeholders collaborated to devise strategies for real-time data collection, which underpinned the planning and execution of the project.

This participatory model, which engaged local leaders and residents, not only facilitated effective data gathering but also empowered citizens to participate in decision-making, both of which are essential elements of smart city development. The Dar Ramani Huria initiative reduced the yearly impact of flooding from fifty thousand (50,000) households to fifteen thousand (15,000) households. Notably, the project helped city authorities save $800,000 annually, which could be used for other interventions.

3.4. Infrastructure Retrofitting

Retrofitting existing infrastructure for efficiency and sustainability is an attractive alternative, offering capital cost savings and promoting urban resilience compared to constructing new infrastructure.

i. Case Study: Solar-Powered LED Street Lighting in Accra, Ghana

Electricity was expensive in Accra, and streetlights routinely went out, especially at night. This situation not only posed security threats but also created economic challenges for the city. In response, the city deployed solar-powered LED streetlights, retrofitting existing poles and light fixtures with solar panels and energy-efficient LEDs. This intervention led to a substantial decrease in electricity costs, improved safety, and reduced maintenance costs. In other words, it not only cut electricity expenses significantly but also enhanced citizen safety and lowered maintenance costs (World Bank, 2019).

Impact Analysis:

- Energy savings: Reduction of municipal electricity bills by approximately 40% annually.
- Maintenance savings: Reduction in maintenance costs by over 60% due to longer-lasting LED bulbs.
- Environmental sustainability: This approach by city authorities promotes environmental sustainability by reducing the dependency on the main grid and reducing carbon emissions.
- Safety impact: Crime incidents at night decreased by approximately 35% in retrofitted areas.

The impact of the solar-powered street lighting in Accra, Ghana, is shown in the table below:

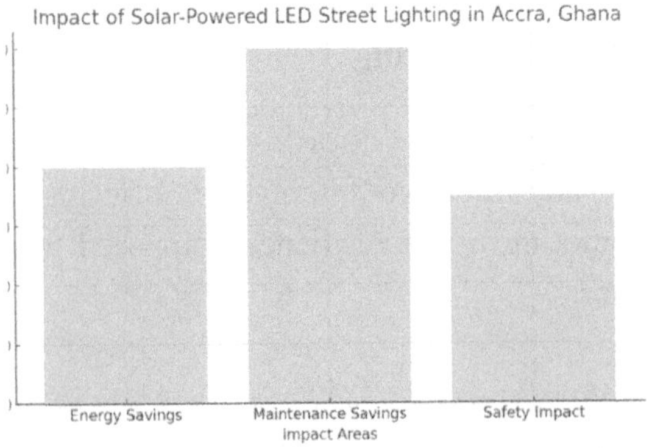

Figure 3.1: The Impact of Solar-powered Street Lighting in Accra, Ghana - *Source: World Bank (2019)*

ii. Summary and Relevance

In Accra, over 14,000 LED energy-efficient lighting fixtures have been installed along 440 km of city streets, replacing outdated mercury and sodium bulbs. This project, highlighted by the World Bank, demonstrates a clear link to the Smart City concept. The solar-powered street lighting promotes environmental sustainability by reducing dependency on the main grid and lowering carbon emissions, while also being cost-effective. These benefits, along with enhanced public safety, align closely with the principles of smart city initiatives, showcasing a modern approach to urban infrastructure.

3.5. Success in a Shoestring: Real-World Wins

Achievable Smart City transformations are not mere dreams; hundreds of jurisdictions worldwide are implementing strategies and interventions to provide practical, affordable solutions to community challenges. This section presents concrete, evidence-based case studies from developing countries that demonstrate how low-cost solutions can deliver efficient responses to complex urban problems.

Even in the absence of smartphones or internet access, SMS-based solutions have proven effective in addressing urban challenges such as traffic congestion and waste management. These on-the-ground examples provide pragmatic, scalable, affordable, and smart solutions for urban problems. This concept can be replicated by policymakers, urban planners, and community leaders in other developing countries.

i. Case Study: Traffic SMS Alert System in Lagos, Nigeria

The authorities in Lagos, Nigeria, adopted a simple SMS-based system that enabled citizens and traffic officials to report congestion hotspots, addressing the city's notorious traffic congestion. This approach is novel because traditional solutions, such as expanding roads or implementing complex infrastructure, are often costly and time-consuming. The system collects and summarizes data and redistributes real-time traffic updates via SMS to commuters (UN-Habitat, 2020). This approach enabled commuters to make informed route decisions by using alternative paths during peak hours.

The congestion leads to increased travel times, driver frustration and fatigue, decreased air quality, lost productivity, and fuel waste. This concept has proven effective for managing urban issues like traffic congestion, even without the need for smartphones or internet connectivity. This innovative approach enables real-time communication and helps alleviate some of the city's most pressing traffic problems. The initiative is part of the city authority's broader efforts to tackle chronic traffic issues, improve mobility, reduce costs, promote environmental sustainability, and stimulate economic activity. An impact analysis, showing the situation before and after the system's implementation, can be found in Figure 3.2 and Table 3.3 below, detailing the positive changes resulting from this innovative solution.

Implementation and Impact:

Setup and Cost-Effectiveness Minimal infrastructure investment was required, relying solely on existing mobile networks and basic data analysis tools.

The Impact Metrics are narrated below:

- Reduced average commuting times by approximately 25% within three years of operation (2016–2019).
- Over 2,000 congestion reports are processed daily, enhancing commuter decision-making and significantly improving overall traffic flow.

- Commuter satisfaction increased substantially, with more than 80% of system users reporting improved daily travel experiences.

Impact Visualization:

The impact of the Traffic SMS Alert System in Lagos, Nigeria is depicted in the figure below:

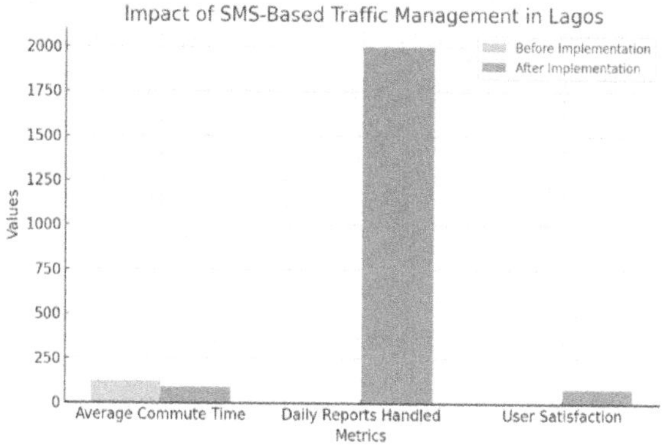

Figure 3.2: Impact Visualization of SMS Traffic Management
Source: World Bank (2020)

Table 3.3: Traffic SMS Alert System in Lagos, Nigeria
Source: UN-Habitat (2020)

Metric	Before SMS System (2015)	After Implementation (2019)
Average Commute Time Reduction (%)	0%	25%
Daily Congestion Reports Handled	0	2,000+
User Satisfaction	N/A	80%

ii. Summary and Relevance

The SMS-based traffic system provides road users with real-time information that significantly enhances transportation management. The SMS traffic concept also aids route planning while helping to reduce congestion and minimize environmental pollution. Through this model, city traffic wardens can gather essential data on traffic flow, congestion levels, and incidents such as car crashes. The real-time data collected is then processed to make adjustments to traffic signals, improving the overall efficiency of the traffic management system. While this approach may not rely on advanced technologies like sensors or the Internet of Things, it exemplifies the effective use of simple technological tools to address urban challenges. This method embodies the principles of a smart city, demonstrating that innovative solutions can emerge from straightforward, accessible technologies. In doing so, it addresses immediate traffic issues and contributes to environmental sustainability, providing a practical and replicable application of smart city concepts.

3.6 Crowdsourced Waste Management Systems

Crowdsourced waste management platforms allow communities to work together at minimal cost and with limited municipal involvement. This concept helps mitigate the challenges posed by waste disposal and its associated health risks. It emphasizes that existing infrastructure can be leveraged for smart city initiatives, even with limited budgets. The benefits of such initiatives are substantial, including resource optimization at minimal cost, promotion of environmental sustainability, and increased community engagement and participation. One notable example is the Community Waste Management Initiative in Pune, India, which demonstrates how localized efforts can enhance cleanliness and sustainability in urban areas, thereby contributing to the broader goal of transforming cities into smarter, more resilient environments.

i. Community Waste Management Initiative in Pune, India

Pune's Solid Waste Collection and Handling (SWaCH) movement serves as an inspiring model for sustainable waste management that alleviates the financial burden on the city. Before SWaCH was initiated, Pune faced challenges such as low recycling rates, high management costs, and significant instances of illegal waste dumping (Chintan, 2019). The SWaCH initiative, supported by local authorities and Non-Governmental Organizations (NGOs), successfully formalized the roles of informal waste collectors and integrated their efforts into the broader municipal waste management framework. This collaboration establishes a direct connection between citizens and the cooperative to optimize waste disposal and recycling processes, significantly reducing reliance on landfills (Chintan, 2019). Through this inclusive approach, SWaCH not only improved urban waste management but also empowered local waste collectors and strengthened community participation and sustainability.

Implementation and Impact:

- Cost and Efficiency: No major infrastructure upgrades were required, as the initiative utilized existing community resources and skills.

Impact Visualization:

The impact of SWaCH initiative in Pune, India is shown below:

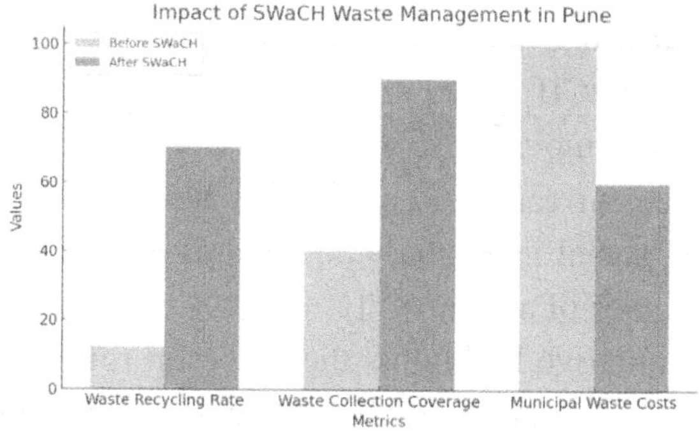

Figure 3.2: The success of the SWaCH initiative in Pune

Source: Chintan (2019).

The SWaCH waste management movement has had a significant positive impact on the community, as mentioned earlier. As shown in Table 3.4 below, the SWaCH initiative achieved an impressive recycling rate of 70%, demonstrating its effectiveness in diverting waste from landfills. Additionally, the movement led to a remarkable reduction in waste management costs, which fell by 40% compared to previous figures.

Furthermore, the SWaCH approach enhanced waste management coverage to an impressive 90%, ensuring that a larger portion of the community is engaged in effective waste disposal and recycling practices. These outcomes not only reflect the SWaCH movement's success but also underscore the importance of community-based initiatives in addressing urban waste challenges.

Table 3.4: The Success Rate of the SWaCH Initiative
Source: Chintan (2019)

Metric	Before SWaCH (2007)	After SWaCH (2017)
Waste Recycling Rate (%)	~12%	70%
Waste Collection Coverage (%)	40%	90%
Municipal Waste Costs	High	Reduced by ~40%
Income Increase for Workers	N/A	Increased by 200%

ii. Summary and Relevance

The SWaCH Cooperative platform in Pune, India, exemplifies how innovative, community-driven approaches to urban waste management can yield significant benefits. By focusing on sustainability, social inclusion, and local empowerment, it reflects the core principles of a Smart City.

This initiative highlights the powerful role of the informal sector in creating smarter, more efficient, and sustainable urban environments. The project's efforts in localized door-to-door collection and recycling have not only improved urban hygiene but also enhanced the health of residents. This promotes healthier populations and a more productive workforce, aligning with the Smart City vision. Community engagement and participation have played crucial roles in reducing municipal waste costs by 40%. SWaCH empowers communities to collaboratively address waste disposal challenges with minimal municipal involvement.

The initiative promotes recycling, transforms waste into valuable resources, and mitigates potential health issues associated with poor waste management. Although the SWaCH concept is not directly related to technology, the model fosters environmental sustainability, social inclusion, data collection, and community engagement which are all key elements of Smart City goals and concepts.

3.7 Informal Settlement Upgrade in Kibera, Nairobi, Kenya

Kibera, one of Africa's largest informal settlements, has historically struggled with inadequate access to clean water, sanitation, and reliable electricity, leading to significant public health concerns and persistent urban poverty. When traditional infrastructure solutions were deemed financially unsustainable, community-based organizations and NGOs implemented small-scale infrastructure improvements through local involvement and low-cost methods (APHRC, 2019).

Faced with the challenge of traditional infrastructure deficits, Community-Based Organizations (CBOs) and Non-Governmental Organizations (NGOs) adopted small-scale infrastructure improvements through local involvement and low-cost methods. These initiatives leveraged community participation and cost-effective strategies for infrastructure enhancement (APHRC, 2019).

Local CBOs and NGOs played a pivotal role by establishing communal sanitation blocks, managed by community cooperatives. This approach quickly led to enhanced hygiene and health outcomes within the settlement. Additionally, simple solar energy installations provided low-cost electricity to households that previously lacked access. These initiatives further improved residents' quality of life and promoted economic empowerment. Overall, these community-driven solutions exemplify the potential of localized interventions to address critical infrastructure gaps in informal settlements like Kibera.

Implementation and Impact:

The impressive outcomes of the Kibera upgrade are depicted in the table below:

Cost-effective Improvements:

Infrastructure projects cost significantly less than municipal-level interventions, largely due to localized planning, volunteer labor, and small-scale installations.

Impact Visualization:

The chart below displays the impressive outcomes of the Kibera settlement upgrade.

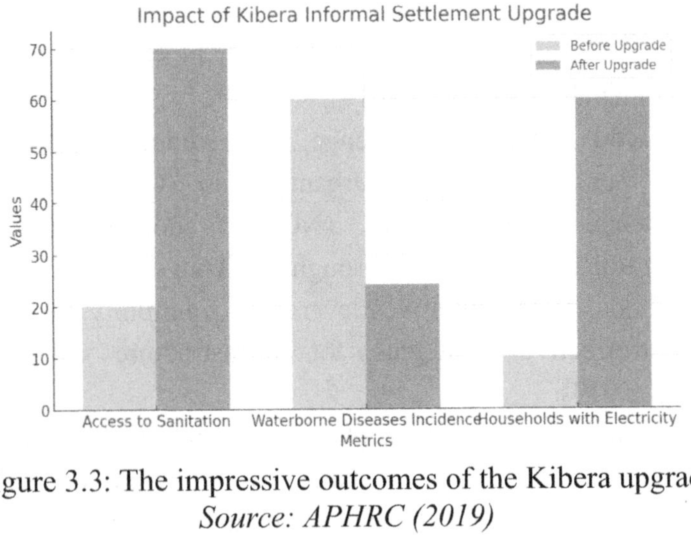

Figure 3.3: The impressive outcomes of the Kibera upgrade
Source: APHRC (2019)

Table 3.5: The success rate of the Kibera upgrade projects
Source: APHRC (2019)

Metric	Before Upgrade (2010)	After Upgrade (2018)
Access to Sanitation (%)	20%	70%
Incidence of Waterborne Diseases (%)	High	Reduced by 60%
Households with Electricity (%)	<10%	>60%

ii. Summary and Relevance

Kibera, located in Nairobi, Kenya, stands as one of Africa's largest and most densely populated informal settlements. The community faces significant challenges, including limited access to clean drinking water, inadequate sanitation, and an unreliable electricity supply. These issues pose serious threats to public health and exacerbate urban poverty. The Kibera project, led by community-based organizations and NGOs, focused on implementing small-scale, low-cost infrastructure improvements.

By engaging in the community and utilizing local resources, the project aimed to bridge the infrastructure gap. Key interventions included building public toilets, installing water kiosks, enhancing sanitation systems, and introducing alternative energy sources such as solar lighting. The results of the Kibera approach were promising. Specifically, incidents of waterborne diseases declined by 60%, and access to sanitation facilities increased by 70%.

This project embodies key elements of a Smart City, as its infrastructure upgrades promoted environmental sustainability and social inclusion. The installation of solar lighting provides the community with a reliable and renewable energy source, while the use of local resources ensures a cost-effective implementation. Although the Kibera project does not rely heavily on advanced technologies, it aligns strongly with Smart City principles by prioritizing affordability, community participation, environmental sustainability, and innovative, locally driven solutions that addressed complex urban challenges.

3.8 Community-Based Flood Mitigation in Jakarta, Indonesia

In Jakarta, a poorly functioning drainage system and rapid urban growth significantly contributed to chronic flooding issues for residents.

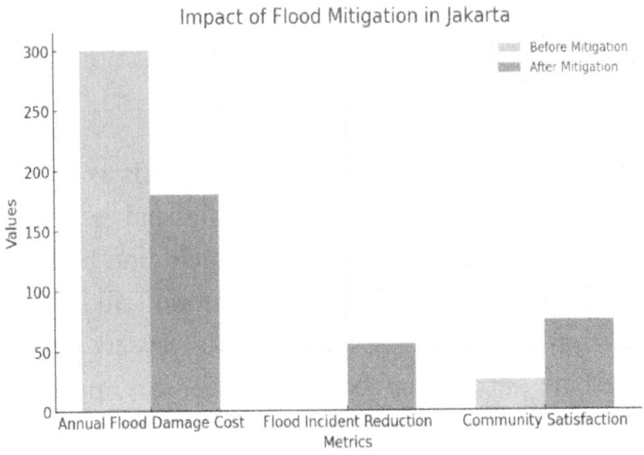

Figure 3.4: Flood mitigation success in Jakarta
Source: World Bank (2021)

Many residents could not afford traditional engineering solutions to tackle the perennial flooding in Jakarta. In response, community groups and local NGOs implemented low-budget, localized flood mitigation initiatives. These initiatives included the construction of drainage systems by community members, the installation of permeable pavement, and the development of flood-absorbent urban gardens (World Bank, 2021).

These grassroots efforts not only addressed the immediate challenge of flooding but also highlighted the potential of local solutions to improve urban resilience and community engagement. By involving residents in the design and implementation of such initiatives, Jakarta is taking significant steps towards more sustainable and effective urban management.

Implementation and Impact:

- Community Engagement: High levels of community involvement reduced project costs dramatically, while also enhancing long-term maintenance and sustainability.

Impact Visualization:

The chart below highlights the flood mitigation success in Jakarta.

Table 3.6: Impact of flood mitigation in Jakarta
Source: World Bank (2021)

Metric	Before (2014)	After Implementation (2020)
Annual Flood Damage Cost	$300 million	$180 million (~40% reduction)
Flood Incident Reduction (%)	0%	55%
Community Satisfaction	Low	High (>75%)

ii. Summary and Relevance

The Jakarta flood mitigation project, which received support from the World Bank, played a pivotal role in moderating flood risk, improving environmental conditions, and enhancing disaster response management. This initiative significantly alleviated the challenges faced by the community, empowered the local council's capacity to handle flood hazards as they arose, and improved local economic activities.

Key to this effort was the collaboration with the Government of the Netherlands and the World Bank in establishing the Flood Management Information System (FMIS). This system empowered Jakarta's local council in various aspects, such as flood assessment, operational planning, maintenance, and monitoring, allowing water to recede more quickly after rainfall into rivers and retention ponds, thereby reducing flood risks. Additionally, the Jakarta flood mitigation project aligns with the Smart City initiative, as the FMIS is essential for effective data collection, analysis, and digital monitoring, facilitating informed decision-making. The project also adopted participatory mapping programs, encouraging community involvement and stakeholder contributions to enhance its effectiveness.

Through local organizations, community members were actively engaged in the decision-making process regarding flood mitigation efforts. According to the World Bank (2012), the costs of the Jakarta flood mitigation project were lower than initially budgeted, which serves as a testament to the value of community engagement. These elements underscore the core principles of the Smart City concept, highlighting the importance of collaboration and technology in urban management.

3.9 Green and Lean: Sustainable solutions for the future

With the growth in population, environmental decline, and climate change, cities are faced with urban challenges that necessitate the adoption of sustainable urban strategies. While many perceive traditional sustainability measures as too expensive, there are numerous low-cost solutions available that can effectively protect natural resources, enhance urban resilience, and foster more favorable climatic conditions. In this section, we will explore practical, low-cost, and sustainable solutions that provide both immediate and long-term economic and environmental benefits.

i. Rainwater Harvesting Systems

The issue of access to clean water remains a significant challenge for urban dwellers, especially in developing countries. One effective solution to this problem is rainwater harvesting, which involves the collection and storage of rainwater for later use. This practice not only reduces dependence on unreliable municipal water systems but also provides a simple and cost-effective method for a sustainable water supply. By implementing rainwater harvesting, communities can enhance their water security at a minimal cost.

ii. Case Study: Rainwater Harvesting in Chennai, India

In response to severe water shortages caused by declining groundwater levels and inconsistent municipal water supply, Chennai implemented a rainwater harvesting mandate in 2003. This program required all buildings to install rainwater harvesting systems, which significantly contributed to groundwater recharge and enhanced urban water sustainability in the city (Narain and Pandey, 2012).

This initiative not only addressed the immediate water scarcity concerns but also promoted a long-term solution by encouraging sustainable practices among residents.

Implementation and Impact:

- Low-Cost Setup: Simple rooftop collection systems and basic storage tanks proved to be cost-effective for individual households and municipal buildings alike.

Impact Metrics:

The effect of harvesting rainwater is shown in the figure below.

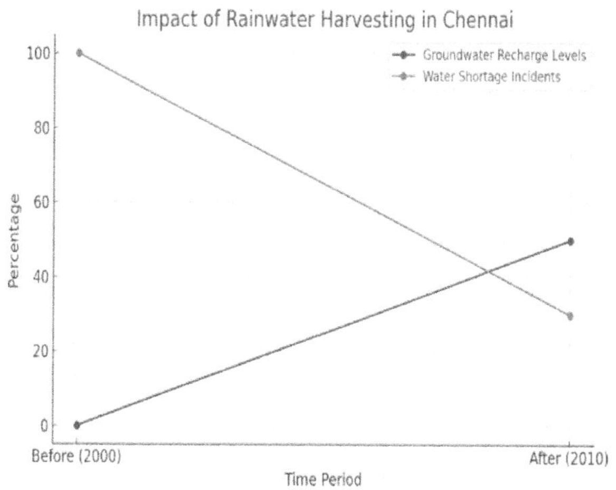

Figure 3.5: Impact of Rainwater Harvesting in Chennai
Source: Narain and Pandey (2012)

The figure above illustrates groundwater levels before and after incidents of water shortages and periods of urban expansion. It clearly shows the trend in rainwater harvesting over time, highlighting how this practice has contributed to improved groundwater recharge levels. This approach has helped to mitigate the perennial water shortages for the residents of the area. However, Table 8 shows how groundwater levels recovered from critically low points to appreciable levels after implementation.

Table 3.7: Impact of Rainwater Harvesting in Chennai, Source: Narain and Pandey (2012)

Metric	Before (2000)	After Implementation (2010)
Groundwater recharge levels	Critically low	Increased by ~50%
Household water expenses (annual)	High	Reduced by ~30%
Water shortage incidents	Frequent	Reduced significantly

iii. Summary and Relevance

The community in Chennai, southern India, faced significant challenges related to inadequate water supply due to rapid urbanization, population growth, and intense pressure on limited natural resources. In response, the Chennai project was initiated, focusing on rainwater harvesting as a sustainable interim solution.

This initiative mandated rainwater harvesting systems for both residential and commercial buildings to facilitate the collection and storage of rainwater, thereby replenishing groundwater levels. By targeting this practice, the project aimed to reduce reliance on alternative water supply sources, ensuring that the community's primary water supply remained sustainable and accessible year-round.

This intervention not only promoted water security but also aligned with the principles of environmental sustainability, which are central to Smart City initiatives. The Chennai project exemplifies sustainability and cost-effectiveness, as community members actively participated in and contributed resources toward the implementation of the project.

3.10 Solar Microgrid System, Affordable and Reliable

Unreliable electricity services make urban living difficult and challenging, especially in informal settlements. This situation necessitated the deployment of a community solar microgrid system by community leaders. The microgrid systems are small, standalone solar-powered energy systems that offer affordable energy solutions to these communities.

This approach not only addressed energy shortages but also empowered residents to take control of their energy sources and consumption. This promotes sustainable development and economic growth within underserved areas. By providing accessible energy, community solar microgrids can stimulate various activities, improve living conditions, and support local businesses, ultimately contributing to the overall advancement of urban environments.

i. Case Study: Solar Microgrids in Dhaka, Bangladesh

In Dhaka, Bangladesh, informal settlements often lacked reliable conventional energy infrastructure, leading to significant challenges for residents. However, the introduction of low-cost solar home systems and community microgrids, installed by Grameen Shakti, a local social business, dramatically improved the quality of life for the Dhaka community. These initiatives have provided access to affordable and sustainable energy solutions, empowered communities, and enhanced their overall living conditions (IFC, 2019).

Implementation and Impact:

- Cost and Maintenance Efficiency: Low upfront costs and minimal ongoing expenses (as solar panels require limited maintenance), ensuring long-term affordability.

Impact Metrics:

The impact of the solar microgrid system is depicted in the figure below:

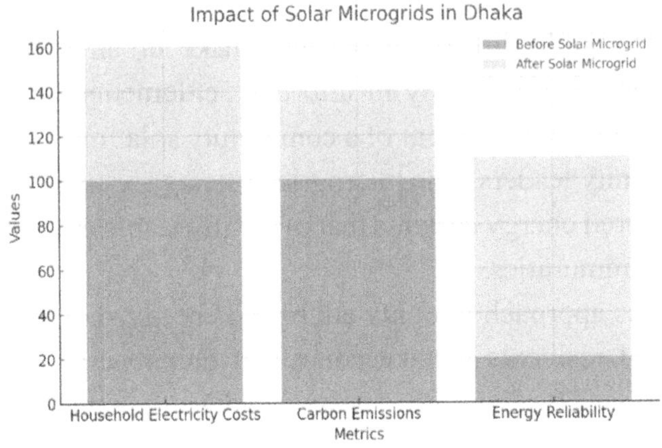

Figure 3.6: Impact of Micro Grids in Dhaka, Bangladesh

Source: IFC (2019)

The figure above depicts not only a reduction in household electricity costs, but also low carbon emissions and efficient energy reliability after the solar microgrid installation.

Table 3.8: Impact of Solar System Installed in Dhaka

Source: IFC (2019)

Metric	Before Solar (2010)	After Solar Microgrid (2018)
Household electricity costs (annual)	High	Reduced by ~40%
Carbon emissions (community level)	High (diesel-based)	Reduced by ~60%
Energy reliability (%)	<20%	>90%

Both the stacked bar chart in Figure 3.7 and Table 3.8 illustrate the positive impact of the solar microgrid deployment on carbon emissions, energy reliability, and efficiency. The data clearly shows a reduction in carbon emissions alongside significant improvements in energy reliability and efficiency following the installation of the solar microgrid (IFC, 2019). These results highlight the effectiveness of renewable energy solutions in promoting sustainability while enhancing overall energy performance.

ii. Summary and Relevance

The microgrid solar lighting system implemented in Dhaka, Bangladesh, has successfully provided a sustainable and efficient energy supply for urban residents. This system aligns with the Smart City concept by incorporating features such as smart control, energy management, and data-driven analytics. Focused on localized energy access, the solar microgrid ensures that energy is reliable, affordable, and sustainable. By enabling local generation of electricity, these microgrids reduced reliance on centralized systems and contributed to a more sustainable environment.

The economic benefits for urban dwellers have been significant, with households experiencing an approximate 40% reduction in annual energy consumption costs, as indicated in the table above. Overall, the project offers considerable cost savings compared to traditional main grid power or diesel-powered generators and provides financial relief for the community.

3.11 Decentralized Water Management Systems

In developing cities, centralized water management systems can often be inefficient and expensive. In contrast, decentralized water systems provide scalable and affordable solutions for managing and distributing water at the local level. These systems enable communities to address their specific water needs more effectively, with enhanced efficiency. This approach reduces the overall costs associated with water supply and management. By empowering local management, decentralized water systems improved water quality and water access for residents.

i. Case Study: Decentralized Water Management in Kigali, Rwanda

The Kigali government, in collaboration with local NGOs, established small-scale decentralized water purification and distribution systems throughout the city of Kigali. These community-managed water systems offered a more cost-effective solution compared to centralized water treatment facilities. The project also ensured high water quality and greater governmental independence (WaterAid, 2020). This approach not only improved access to clean water but also empowered local communities to take charge of their water management processes.

Implementation and Impact:

- Affordable Technology:
  - Small, community-level purification units using basic filtration and chlorination techniques.

Impact Metric: The impact of the project is shown in the figure below:

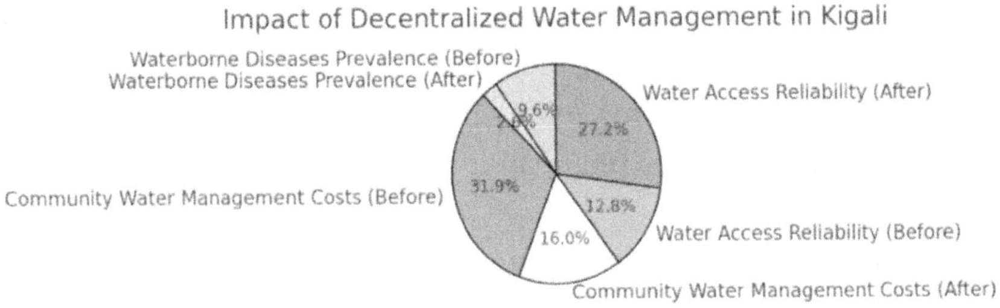

Figure 3.7: Impact of decentralized water management system in Kigali; *Source: WaterAid (2020).*

Figure 3.8 and Table 3.9 below illustrate the positive impact of decentralized water management in Kigali. The data clearly highlights a significant reduction in waterborne diseases and an improvement in the supply of reliable water following the implementation of the projects. These results underscore the effectiveness of decentralized water management systems, which help in enhancing public health and ensuring a consistent water supply for the community.

Table 3.9: Impact of decentralized water system in Kigali, Source: WaterAid (2020)

Metric	Before Implementation (2012)	After Implementation (2019)
Waterborne diseases prevalence (%)	High (~30% of population)	Reduced to ~8%
Community water management costs	High	Reduced by ~50%
Household water access reliability (%)	Low (40%)	High (>85%)

ii. Summary and Relevance

The Decentralized Water Management in Kigali (DWMSK) initiative enabled citizen control over their water resources, and it enhanced leak detection and repair capabilities while improving oversight. The project primarily focused on several critical areas, including managing water consumption, detecting and repairing leaks, inspecting and calibrating meters, replacing water service pipes, and providing training to build the capacity of local personnel. The DWMSK exemplified a problem-driven approach by concentrating on improved water delivery.

This approach helped to reduce leakages and losses while empowering local technical teams to adopt proactive and innovative solutions. Following its implementation, the initiative yielded significant results: community water management costs were reduced by approximately 50%, the incidence of waterborne diseases dropped to about 8%, and the reliability of household water access improved dramatically from a low of 40% to over 85%. By enhancing the technical skills of local community members to tackle these challenges, the project not only had a positive impact on the budget but also laid a strong foundation for future scalability. Even though the DWMSK did not rely heavily on high-tech solutions, its solid fundamentals underscore the potential for smarter and more resilient urban water management strategies.

3.12 Beyond Gadgets: Non-Tech Paths to Smarter Cities

Technology can drive significant advancements in urban areas, but it is not the only factor essential for achieving smart urban development. Many cities in developing countries have successfully implemented non-technological initiatives that emphasized community participation, social innovation, and the smart use of resources.

By harnessing local knowledge, strengthening community engagement, and promoting participatory governance, these initiatives have delivered remarkable results with minimal budgets and investments. While the transformation of communities into smart cities heavily relies on technology, it is crucial to recognize that technology alone is not a definitive solution. Several developing countries have demonstrated success in championing smart city initiatives without heavy reliance on tech.

These countries instead focused on community engagement, knowledge sharing, inclusive governance, and resource pooling. These processes can be considered "smart" in their own right, as they promote locally relevant solutions. They also cultivate community ownership and enhance trust, which are conditions for building resilient urban environments.

i. Participatory Urban Planning

A participatory planning process that brings community members into the urban decision-making process helps to alleviate some urban challenges. This approach empowers communities to prioritize resources and execute community projects on a reduced budget. Not only does this reduce project costs, but it also enhances project efficiency, sustainability, and community ownership (Mitlin and Bartlett, 2018). With participatory urban planning, stakeholders, local citizens, and government officials play contributory roles in community projects. Hence, these projects are executed at a lower cost. This participatory initiative is one of the key branches of the Smart City concept.

ii. Case Study: Urban Renewal through Participatory Planning in Medellín, Colombia

Once infamous for crime and social inequality, Medellín's radical transformation into a model of urban renewal is largely attributed to the principles of participatory planning. Beginning in the early 2000s, initiatives like the Urban Integral Project actively involved residents from the poorest neighborhoods. The residents participated in the design of public spaces, parks, and community facilities. Rather than imposing solutions from outside, city authorities worked collaboratively with residents, significantly enhancing the projects' social relevance and sustainability (Blanco and Kobayashi, 2009). While this participatory approach to urban renewal aligns with smart city development principles, it does not rely heavily on technological integration.

Key Outcomes:

- Reduced crime rates by 40% in participating neighborhoods (2004–2015).
- Increased community participation in municipal governance, with resident engagement in community projects rising significantly.
- Enhanced social cohesion and trust between citizens and municipal authorities (Blanco and Kobayashi, 2009).

Expert Insight:

"Participatory planning made Medellín's urban renewal successful because it empowered residents as partners in decision-making, rather than as passive recipients of imposed solutions."

iii. Community-Based Waste-to-Energy Systems

At the community level, converting waste into energy provides an effective solution to the waste crisis confronting residents of Medellín. This approach offers the community a cost-effective and renewable energy alternative. Utilizing technologies from outside the traditional tech sector, such as small-scale biodigesters, proves particularly beneficial in urban areas where municipal waste collection services may be lacking. These technologies not only help manage waste more effectively but also promote environmental sustainability and enhance local energy security.

iv. Case Study: Bio-Digester Initiatives in Kampala, Uganda

In Kampala, urbanization had a significant negative impact on waste management, coupled with inadequate municipal infrastructure for solid waste management. To address these issues, a network of sustainable biodigesters was built across various informal settlements. These biodigesters reuse organic waste to produce biogas for household cooking and heating, thereby reducing dependence on fossil fuels and alleviating the pressure on waste management systems (Practical Action, 2020). Moreover, this innovative system is accessible via any mobile device, making it even more user-friendly and efficient.

Impacts Observed:

- Significant reduction in organic waste sent to landfills (up to 60% in pilot communities).
- Annual household fuel costs decreased by approximately 30%.
- Reduction in household air pollution from traditional biomass cooking methods.

v. Community-Led Transport Networks

Infrastructure investment alone doesn't guarantee optimal

results for transport networks; community-led initiatives also play a crucial role in improving urban mobility. Integrating informal transport options into more formal transportation networks further enhances mobility and accessibility. This collaborative approach enhances accessibility and efficiency in urban transit and demonstrates the value of community involvement in transportation planning.

vi. Case Study: Bus Rapid Transit (BRT) in Curitiba, Brazil

The successful implementation of the Bus Rapid Transit (BRT) system in Curitiba, Brazil, exemplifies an innovative solution to public transportation challenges. This system offers high-quality transportation services without the need for expensive transportation infrastructure.

Table 3.10: Cost Comparison of Transport Solutions in Curitiba
Source: Cervero (2013)

Transport System	Initial Investment Cost (per km)	Annual Operational Cost	Commuter Satisfaction (%)
Light Rail System	~$40 million	High	60%
Community-Led BRT System	~$1.5 million	Moderate	85%

A key factor in this transformation was community engagement during the planning stages, as well as the strategic use of the existing informal minibus network. By incorporating these routes into the BRT grid, the system not only helps to alleviate congestion in the city but also optimizes capital investments (Cervero, 2013).

Outcomes:

- Reduced average commute time by 30% within the first five years.

- Decreased reliance on private vehicles, leading to reduced emissions and traffic congestion.
- Achieved high rider satisfaction, with over 70% of residents using public transport regularly.

vii. Summary and Relevance

Smart Cities emphasize the importance of inclusive development and community empowerment. Engaging local communities in the decision-making process, implementation, and maintenance of initiatives like bio-digesters not only creates jobs but also strengthens economic empowerment.

This collaborative approach promotes environmental sustainability by enabling locals to convert waste into clean and affordable energy. With biogas now being used for heating and cooking, there is a reduced reliance on wood for fuel. This reduces health-related issues as a result of smoke exposure and helps to mitigate environmental pollution. However, the concept also provides sustainable income for the community.

The biogas initiative was implemented with a modest budget, and it yielded positive results because of community involvement and inclusive participation of all stakeholders. Although this approach may not have a direct link to advanced technological infrastructure, it has potential for future integration within the Smart City framework. The system's accessibility via mobile devices allows for data collection and analysis, features typical of Smart City applications. The BRT system enhances urban mobility while reducing traffic congestion through its consultative approach, community engagement, and well-thought-out implementation strategy.

It features dedicated bus lanes and high-capacity buses with an integrated fare collection system. What sets the Curitiba BRT system apart is its innovative integration of informal minibuses into the formal transit system. This inclusion improved mobility, coverage, and efficiency. This approach ultimately reduced the initial project cost of $1.5 million to a moderate figure, with approximately 85% user satisfaction. The Curitiba BRT system contributed significantly to reducing congestion and pollution, while enhancing mobility for all citizens.

3.13 Pooling Resources: Partnerships That Pay Off

In a resource-constrained environment, financial limitations often restrict the scope and scale of urban development projects. However, strategically pooling resources through Public-Private Partnerships (PPPs) and Community-Based Funding (CBF) mechanisms can offer viable alternatives. These approaches can bring in additional resources, foster greater development, share knowledge and risks, and introduce innovative ideas. PPPs, in particular, extend access to resources and capabilities that surpass those of conventional funding sources.

This section serves as a comprehensive guide to various partnership models and their documented successful examples. It also illustrates how cities can overcome financial barriers to achieve meaningful urban development. By leveraging these models, municipalities can enhance their ability to implement substantive changes that benefit their communities.

i. Public-Private Partnerships (PPPs)

Public-Private Partnerships (PPPs) are collaborative agreements between government agencies and private-sector companies aimed at investing in and sharing the costs, risks, and responsibilities. PPPs are associated with delivering public goods or services, as well as with building and maintaining infrastructure.

These partnerships are particularly valuable for cities operating under tight municipal budgets. Additional resources for infrastructure development can be procured through PPPs to improve the livability of citizens without burdening the central government (World Bank, 2021). Through this model, cities can benefit from the expertise and efficiency that private-sector partners bring on board, including access to upfront capital.

Key Benefits of PPPs:

- Shared Financial Risk: PPPs help mitigate financial risks by dividing the costs and responsibilities between the public and private sectors.
- Access to Expertise and Innovation: Private companies often bring cutting-edge technology and more efficient management processes and enhance the quality of services and projects.
- Long-term Financial Sustainability: By tying payment structures to service delivery and performance.
- Infrastructure Sustainability: PPPs ensure that infrastructure projects remain sustainable and deliver value for money over time.

ii. Case Study: Public Wi-Fi Initiative in Kigali, Rwanda (AM HERE)

Kigali aimed to enhance internet access for its residents but faced significant budgetary constraints. In response to this challenge, the city formed a Public-Private Partnership (PPP) with leading telecommunications companies. These private firms invested in the required infrastructure to support widespread connectivity. The city government provided essential regulatory support and strategic incentives to facilitate this initiative. This collaborative effort enabled the rollout of massive public Wi-Fi, which greatly contributed to digital inclusion within the community (UN-Habitat, 2019).

Implementation and Impact:

- Cost-sharing: The partnership reduced the financial burden on municipal authorities by over 80%, with private companies covering upfront infrastructure costs.

Impact Metrics:

The impact of the Public Wi-Fi Initiative in Kigali is captured in the table below:

Table 3.11: Impact of PPP on the Deployment of Wi-Fi in Kigali
Source: UN-Habitat (2019)

Metric	Before PPP (2015)	After PPP (2020)
Internet Accessibility (%)	15%	70%
Municipal Expenditure (project)	~$10 million (estimated)	~$2 million
User Satisfaction	Low	High (~85%)

iii. Summary and Relevance

The Public-Private Partnerships (PPPs) model comes with several benefits, including spreading costs and risks, while promoting digital inclusion. PPPs enhance critical infrastructure development without straining municipal budgets. A notable example is the Public Wi-Fi Initiative in Kigali, which was part of the Smart Kigali program.

This initiative successfully provided free public Wi-Fi in various key locations, including bus terminals, municipal markets, hospitals, the airport, public transport hubs, schools, and libraries. The project was a collaborative effort among several government agencies, such as the Rwanda Utilities Regulatory Authority (RURA), the Rwanda Development Board (RDB), the Ministry of Youth and ICT, the City of Kigali, and private telecom operators.

As a result of this initiative, internet access in Kigali increased from just 15% to an impressive 70%. Additionally, the municipal infrastructure costs dropped significantly, from an estimated $10 million to only $2 million, while user satisfaction ratings soared to 85%, reflecting a substantial improvement from previous low levels.

The PPP model was particularly effective in this context, as it combined public sector resources with private-sector expertise, leading to enhanced service delivery and infrastructure development.

3.14 Community-Based Funding Models

Community-based funding models effectively mobilize financial resources from residents, local businesses, NGOs, and other stakeholders. By leveraging local commitment, knowledge, and resources, these models ensure that projects are closely aligned with the specific needs of the community. This alignment not only enhances the initiative's sustainability but also increases its efficiency.

This is because community members are more likely to support and engage with projects that directly address their concerns and aspirations. Ultimately, these models foster a sense of ownership and responsibility within the community, leading to more successful and enduring outcomes.

i. Case Study: Cooperative Infrastructure Development in Nairobi, Kenya

Municipal funding alone has never been sufficient to meet the basic infrastructure needs of Nairobi's informal settlements. To address these challenges, residents formed federated community cooperatives that collaborated with local NGOs and private donors to pool resources. This collective effort enabled them to construct essential infrastructure, including sanitation blocks, water distribution points, and road improvements. Nevertheless, members utilized locally sourced labor and materials for the construction of the projects. Their efforts yielded significant reductions in both implementation timelines and project costs (APHRC, 2019). This cooperative approach highlights the power of community engagement in addressing infrastructure deficits and in improving living conditions in informal settlements.

Implementation and Impact:

- Cost and Efficiency: By using community labor and locally sourced materials, project costs were reduced by nearly 60% compared to municipally managed projects.

Impact Metrics:

The Municipal Project cost in Nairobi is shown in the table below:

Table 3.12: Municipal Project cost in Nairobi, Kenya
Source: APHRC (2019)

Metric	Municipal-led Project (Cost)	Community-led Project (Cost)
Sanitation Block per Unit	~$25,000	~$10,000
Water Distribution Points	~$20,000	~$8,000
Implementation Time	1 year	4 months

ii. Summary and Relevance

The Cooperative Infrastructure Development initiative in Nairobi, Kenya, seeks to enhance water delivery by making it more cost-effective, efficient, and sustainable. This model illustrates how community-driven funding can substantially lower project costs and adhere to timelines while delivering relevant local solutions. The effectiveness of the community-based funding approach is further enhanced when local stakeholders are actively engaged in the process.

This cooperative model not only fosters cost savings but also helps to eliminate operational inefficiencies. Through community-led infrastructure development, Nairobi can work towards achieving smart city status by integrating technology, improving data collection, and enhancing data management practices. Integrating community engagement and innovative solutions empowers city authorities to better address their infrastructure needs sustainably.

3.15 Leveraging International Development Funding

Many international development organizations and multilateral financial institutions regularly offer grants, loans, and technical assistance for sustainable urban projects. Cities that adopt locally grown initiatives that are sustainable, resilient, and focused on poverty reduction are more likely to attract significant funding from these international entities. Cities that demonstrate commitment to these shared goals are better positioned to secure funding for their projects.

i. Case Study: Green City Initiative in Addis Ababa, Ethiopia

Addis Ababa is on track to become the first green city in Africa and has successfully secured significant international funding. The funding will help Addis Ababa expand its infrastructure projects in line with the Sustainable Development Goals (SDGs). By focusing on sustainable, community-oriented initiatives, the city has attracted interest from various international agencies, including the World Bank, Deutsche Gesellschaft für Internationale Zusammenarbeit (GIZ), and the United Nations Development Programme (UNDP).

The allocated funding has been directed toward a wide range of strategic projects, such as improving energy-efficient and affordable housing, expanding urban green spaces, upgrading public transportation systems, and developing integrated information and monitoring systems. These investments aim not only to reduce environmental impacts but also to enhance the quality of life for residents by promoting cleaner air, better mobility, and more inclusive urban services.

As a result, Addis Ababa has been able to strengthen its urban resilience, improve service delivery, and build institutional capacity, all while maintaining a manageable financial burden on its annual budget (World Bank, 2022). This balanced and forward-looking approach illustrates how cities in developing regions can leverage international partnerships and targeted investments to achieve sustainable urban transformation without compromising fiscal stability.

Implementation and Impact:

- Attractiveness of Initiative: The alignment with SDG 11 (Sustainable Cities and Communities) was instrumental in attracting diverse international funding sources.

Impact Metrics

The analysis of international development funds on sustainable projects is demonstrated in the table below:

Table 3.13: Impacts of International Development Funds on Projects, Source: World Bank (2022)

Metric	Before Funding (2015)	After Funding (2022)
Housing Affordability	Low	Increased significantly
Urban Green Spaces (sq. km)	10 sq. km	35 sq. km
Sustainable Transport Coverage (%)	<20%	>60%

ii. Summary and Relevance

The above case illustrates the significant potential and impact of international development funding on sustainable city projects. Collaborations among public and private sectors, community-based funders, and participants in international markets can leverage their unique strengths to mobilize resources. This will build the capacity of community team leaders to execute underfunded urban infrastructure projects. These collaborative approaches not only facilitate funding but also promote the sharing of knowledge, talent, and community involvement. Ultimately, such projects foster economic sustainability, enhance livability, and improve the overall effectiveness, impact, and longevity of initiatives. By enabling cities to achieve large-scale urban transformations without straining budgets, these partnerships provide a feasible pathway toward creating affordable smart city solutions. By leveraging resources from diverse providers, cities gain access to innovative infrastructure solutions, allowing them to address the challenges posed by growing populations while becoming more environmentally and socially responsible.

3.16 Conclusions

In this chapter, we explored the concept of affordable Smart Cities, challenging the common belief that innovative urban development requires significant investments in advanced technologies and infrastructure. We highlighted how low-tech, community-driven solutions can substantially enhance urban living, particularly in resource-constrained areas.

By showcasing practical innovations, ranging from citizen engagement and mobile applications to open-sourced mapping and infrastructure retrofits, this chapter demonstrated that meaningful urban transformation is possible even on a limited budget. We emphasized the importance of leveraging local knowledge, fostering purposeful partnerships, and creatively applying technology to improve service delivery, governance, and quality of life. These were executed without the need for hefty financial resources. Examples such as community-based waste-to-energy systems and sustainable participatory planning models illustrate the effectiveness of non-tech solutions.

Historically, urban redevelopment often seemed unattainable due to an over-reliance on government funding and substantial capital requirements. Furthermore, this chapter also underscored various environmental sustainability models by introducing green practices, including rainwater harvesting, solar microgrid initiatives, and decentralized water management systems.

These examples affirmed the fact that environmentally friendly projects do not necessarily entail a high financial burden. Lastly, we examined how cities can capitalize on shared resources through Public-Private Partnerships (PPPs) and Community-Based Funding models.

These approaches offer financial leverage, access to expertise, and innovative solutions. Cities like Kigali, Nairobi, and Addis Ababa are clear examples of how effectively they have utilized partnerships to advance urban development agendas without straining municipal budgets. In summary, this chapter illustrates that Smart Cities can thrive on low-tech, low-cost principles while providing scalable, locally driven solutions that enable sustainable and environmentally friendly urban transformation.

References

Angelidou, M. (2015). Smart Cities: A conjuncture of four forces. *Cities, 47*, 95–106. https://doi.org/10.1016/j.cities.2015.04.008

APHRC. (2019). *Nairobi Cross-sectional Slum Survey (NCSS)*. African Population and Health Research Center, Nairobi, Kenya.

APHRC. (2019). *Community-Based Infrastructure Development in Nairobi's Informal Settlements*. Nairobi, Kenya: APHRC.

Blanco, C. & Kobayashi, H. (2009). Urban transformation in slum districts through public space generation and cable transportation at Northeastern Area: Medellín, Colombia. *Journal of International Social Research, 2*(8), 75–90.

Cervero, R. (2013). Bus rapid transit (BRT): An efficient and competitive mode of public transport. *Institute of Urban and Regional Development (IURD), Berkeley.*

Chintan Environmental Research and Action Group. (2019). *Informal Waste Sector and Waste Pickers: Pune's Experience*. Chintan Report.

Chintan Environmental Research and Action Group. (2019). *Informal Waste Sector: Pune Case Study*. Chintan Publication.

GIZ. (2018). *Digital Solutions for Sustainable Urban Development*. Deutsche Gesellschaft für Internationale Zusammenarbeit (GIZ).

Güneralp, B., S. Lwasa, H. Masundire, S. Parnell, and K. C. Seto. (2017). "Urbanization in Africa: Challenges and Opportunities for Conservation." Environmental Research Letters 13 (1): 015002.

Iliffe, J. et al. (2017). Mapping for disaster management: Dar Ramani Huria project. *Tanzania.*

IFC (International Finance Corporation). (2019). *Solar Microgrids: Scaling up Solar Energy Access in Bangladesh*. World Bank Group.

McKinsey Global Institute. (2018). *Smart cities: Digital solutions for urban transformation*. McKinsey & Company.

Mitlin, D., & Bartlett, S. (2018). Co-production – key ideas.

Environment and Urbanization, 30(2), 355–366. https://doi.org/10.1177/0956247818781457

Okolloh, O. (2009). Ushahidi, or 'testimony': Web 2.0 tools for crowdsourcing crisis information. *Participatory Learning and Action, 59*(1), 65-70.

OECD/SWAC. 2020. Africa's Urbanization Dynamics 2020: Africapolis, Mapping a New Urban Geography. Paris: OECD Publishing

Practical Action. (2020). *Urban Waste Management in Uganda: Transforming Challenges into Opportunities*. Practical Action Publications.

WaterAid. (2020). *Decentralized Water Management: Kigali Case Study*. WaterAid Rwanda Country Program.

UN-Habitat. (2019). *Leveraging Public-Private Partnerships for Smart Urban Development: Kigali Case Study*. United Nations Human Settlements Programme.

UN-Habitat. (2020). *Smart Cities and Mobility: Leveraging Technologies for Sustainable Transportation*. United Nations Human Settlements Programme.

Uyarra, E., Zabala-Iturriagagoitia, J. M., Flanagan, K., & Magro, E. (2020). Public procurement, innovation and industrial policy: Rationales, roles, capabilities and implementation. Research Policy, 49(1), Article 103844. Https: //doi.org/10.1016/j. respol.2019.103844

World Bank. (2019). *Sustainable Urban Energy and Emissions Planning*. World Bank Group.

World Bank. (2020). *Smart and Sustainable Transport Systems in Lagos*. Urban Development Department, World Bank Group.

World Bank. (2021). *Public-Private Partnerships for Urban Development: Best Practices Guide*. Washington, DC: World Bank Group.

World Bank. (2022). *Green Cities in Africa: Addis Ababa's Urban Sustainability Transformation*. Washington, DC: World Bank Group.

CHAPTER 4
A Framework for Sustainable and Cost-Effective Smart Cities

4.1 Introduction

A crucial problem as cities around the world develop into smarter, more sustainable urban centers is striking a balance between affordability and innovation. The goal of cost-effective smart city solutions is to optimize resources, cut costs, and promote sustainable growth while building infrastructure that is ready for the future. These methods aim to maximize value in urban development rather than merely reducing expenses. It is an innovative city that leverages information and communication technologies (ICT) and other tools to enhance competitiveness, quality of life, and the effectiveness of urban operations and services.

It is known as a smart sustainable city if it meets the economic, social, environmental, and cultural demands of current and future generations. To solve the problems associated with urbanization and improve quality of life, a framework for sustainable and economically viable smart cities integrates technology, government, and citizen participation. To attain long-term benefits, this strategy emphasizes a holistic view of urban development, combining social, economic, and environmental sustainability.

This entails embarking upon solving environmental issues, advancing social justice, and guaranteeing long-term economic sustainability. It requires allocating resources as efficiently as possible, making effective use of technology, and ensuring that initiatives are both profitable and accessible to all parties involved.

4.2 Theoretical Framework

A Smart City framework is made up of many different elements, such as technology, governance, citizen involvement, and sustainability (Gowda et al., 2023). It is designed to improve the quality of life for its citizens while addressing the complex problems that urbanization has brought about. Consequently, a structured approach to addressing environmental, social, and economic issues in a sustainable manner is provided by sustainability frameworks and standards, which are established guidelines and requirements.

In order to enhance the quality of life while tackling urban issues, a sustainable and economical smart city framework should integrate technology, governance, public involvement, and sustainability. This paradigm, which emphasizes environmental, economic, and social sustainability, ought to be based on Elkington's (1994) Triple Bottom Line (TBL) theory.

This theory evaluates how companies impact people, the environment, and profits to determine their overall sustainability and social responsibility. By taking a holistic approach, it encourages businesses to consider social and environmental factors alongside financial performance, helping to create long-term value and a more sustainable future for all stakeholders.

As a corporate framework, it measures success across three essential dimensions: profit, which reflects financial performance; people, which addresses community impact, social responsibility, and ethical labor practices; and planet, which focuses on reducing carbon emissions, conserving resources, and advancing environmental sustainability.

As illustrated in Figure 4.1, Salawu (2024) highlighted this in a paper presentation aimed at eradicating poverty through coastal economic growth during an international conference in Vietnam, themed "Digital Innovation: Sustainability through the Lens of the Sustainable Development Goals."

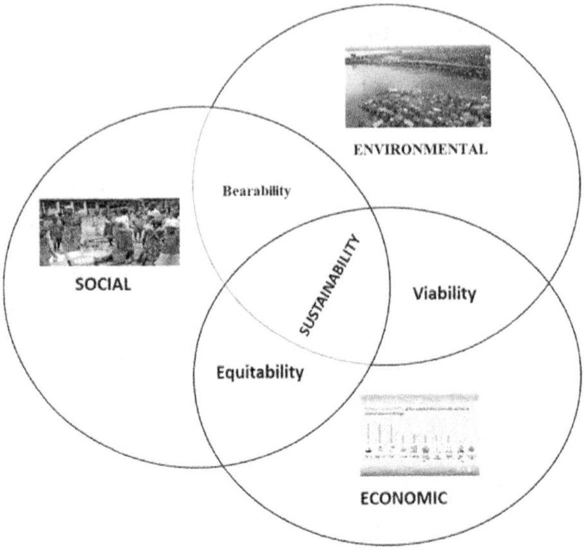

Figure 4.1: Sustainability Concept (Salawu, 2024)

Digital twins, blockchain, IoT, and artificial intelligence are examples of emerging technologies that can be extremely important in accomplishing these objectives. Cities can improve public services, maximize resource management, and encourage community involvement by utilizing these technologies, ensuring that all opinions are heard during the decision-making process. In the end, this all-encompassing strategy not only stimulates creativity but also creates resilient urban environments that can adapt to new challenges.

4.3 The Four-Stage Smart City Framework

To understand and apply the Framework of Urban Sustainability and Cost-Effectiveness, this chapter presents a four-stage approach, which includes the following steps:

- Diagnosing the current state of the city
- Defining a vision for change and setting priorities

- Determining how to finance the plan that achieves and demonstrates fiscal sustainability
- Monitoring and evaluation.

This structured methodology not only aids in identifying the particular challenges that urban areas face but also encourages community engagement and collaboration. By adhering to these stages, cities can successfully implement sustainable practices that improve the quality of life for their citizens while conserving resources for future generations. Through the involvement of all stakeholders in the decision-making process, this comprehensive approach guarantees more inclusive and equitable results.

In the end, a dedication to sustainability will enable communities to prosper and adapt to a constantly shifting environment. Urban planners can foster trust among locals and motivate them to actively participate in community development by emphasizing openness and transparency. In addition to fortifying ties within the community, this collaborative spirit fosters a sense of shared accountability for environmental stewardship. Furthermore, this chapter explores tactics and tools that can assist cities in achieving their smart transformation objectives at a reasonable cost.

Urban planners may develop more effective systems that improve public services by utilizing affordable solutions like open-source software, data analytics, and community involvement. Public-private partnerships with local companies and organizations can also provide valuable resources and expertise to support these projects.

4.4 Key Components of the Framework

Strong infrastructure, effective governance, active citizen participation, and technological innovation are all key elements of a sustainable and cost-efficient Smart City framework. Together, these components address urban challenges, improve quality of life, and promote sustainable development. This chapter focuses on how the framework is built around four pillars including environment, society, governance, and economy to help readers better understand its full potential. Policymakers and urban planners can use this framework as a tool to assess the effectiveness of Smart City initiatives in real-world applications.

Vision and Strategy Development

It is important to create a clear vision for the Smart City that defines precise goals and devising a complete strategy to attain them. It is essential to ensure that all stakeholders are aligned and work toward common goals by encouraging collaboration among government agencies, business sector collaborators, and the community. In order to effectively respond to evolving challenges and opportunities, the strategy must be assessed and adjusted continually.

Smart Infrastructure and Technology Integration

Improving urban services and infrastructure through the use of IoT, smart grids, and AI technology has the potential to enhance resource management, public safety, and citizens' quality of life. Furthermore, data gathered by these technologies can help inform future urban planning and development initiatives.

Data-Driven Governance

Data analytics and machine learning enable better decision-making, increased productivity, and enhanced transparency. In the end, improved resource management and greater accountability results from enterprises' capacity to see trends, forecast outcomes, and optimize processes thanks to these cutting-edge technologies. By utilizing these resources, organizations can promote an innovative and continually improving culture.

Environmental Sustainability

It is imperative to concentrate on cutting emissions, supporting renewable energy, and conserving resources through sustainable practices in order to make the planet healthier and ensure a higher standard of living for future generations. Adopting these values will allow us to preserve the environment while promoting innovation and economic expansion.

Economic and Social Inclusivity

A cohesive and vibrant urban environment depends on ensuring that all citizens benefit from smart city initiatives, addressing social inequalities, and promoting economic opportunities. By prioritizing inclusiveness in the planning and execution of these initiatives, cities can develop sustainable solutions that benefit everyone in the community.

Resilience and Security

A smart city that is resilient, secure, sustainable, and economical makes use of social networks, infrastructure, and technology to improve economic stability, environmental sustainability, and catastrophe preparedness while lowering costs. This entails combining a number of tactics, such as resource management, community involvement, and wise governance, to create a city that can recover from and adapt to shocks and stressors, both natural and socioeconomic. Cities can improve the security of their residents and infrastructure by tackling the threats posed by cyberattacks, natural disasters, and other risks. In order to guarantee the safety and resilience of urban settings against a variety of challenges, proactive measures can be put in place to defend the city from hazards such as cyberattacks and natural disasters.

Governance and Collaboration

A sense of community ownership over local initiatives can be fostered, and decision-making can be made more transparent by implementing tactics such as establishing efficient governance structures, encouraging public-private partnerships, and including residents in the planning process. Ensuring that a range of opinions are heard will enable the development of creative solutions that are also suited to the particular requirements of the community.

This cooperative approach between the public and governmental entities fosters responsibility and trust, increasing the likelihood of successful implementation and long-term sustainability when local programs inspire individuals to feel invested in them.

Citizen Engagement

More sustainable and efficient solutions may result from involving locals in the design and implementation of smart city initiatives to ensure that their needs and concerns are addressed. Involving the community ensures that the projects accurately represent the needs and goals of the local population and promotes a sense of ownership. This can be achieved by engaging citizens through the following methods:

- Talk about ways to boost community involvement in smart city projects, like public forums or workshops.
- Examine how smart technology might improve accountability and transparency in public services.
- Emphasize effective case studies where local residents' participation resulted in creative urban sustainability solutions.
- Analyze how collaborative governance affects policymaking and how crucial it is for resolving local issues.
- To encourage continued communication and collaboration, provide methods for fostering trust between locals and government representatives.

The pursuit of creating smart cities that are both economical and sustainable requires a strategic framework that encourages cooperation, creativity, and innovation. Cities may adapt to the challenges of urbanization and enhance the quality of life for their citizens by addressing the interrelated aspects of technology, social justice, environmental impact, and economic efficiency.

By adopting flexibility and stakeholder engagement, the framework offers a method for cities to develop into smarter, more sustainable communities that benefit future generations. In order to support a broader trend towards global sustainable urban development, as well as the success of specific projects, collaborative efforts and the integration of diverse ideas are crucial.

The dedication to developing smart cities has the potential to revolutionize urban living in a variety of places, such as Lagos, Nigeria, where tailored frameworks can meet regional requirements. Through the implementation of this comprehensive strategy, cities worldwide can strive to create environments that foster resilience, sustainability, and prosperity for all residents.

4.5 Implementation Strategies of Sustainable and Cost-Effective Smart City

A framework for creating Smart Cities that use information and communication technology (ICT) to achieve cost-effectiveness, and sustainability is likely depicted in Figure 4.2. To increase productivity, lessen the impact on the environment, and improve the quality of life for locals, this framework would involve integrating ICT with urban infrastructure and services.

The framework can help create creative solutions that are suited to the particular difficulties that every city faces by encouraging cooperation between different stakeholders, such as public and corporate organizations, as well as local communities. In the end, this all-encompassing strategy seeks to build an urban environment that is resilient enough to adapt to changing needs and technological breakthroughs. A civilization dedicated to cultural sustainability is depicted in Figure 4.2, which emphasizes intergenerational justice and the connections between social well-being, economic viability, and environmental stewardship. This means finding a balance between the demands of the economy, people, and the environment. Everyone must have a high standard of living, and natural resources must be protected for future generations. A sustainable society essentially recognizes that the health of the world and human well-being are inextricably linked.

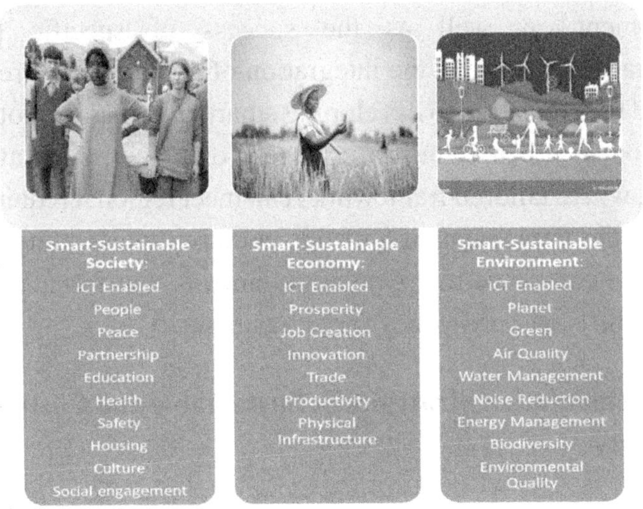

Figure 4.2: A framework for sustainable and cost-effective Smart Cities, *Source: Author (2025)*

It highlights how important it is to coexist peacefully with the environment to guarantee the well-being of future generations. However, some contend that if data collection grows more widespread, such integration may result in people's privacy concerns and increasing surveillance. Furthermore, the dependence on technology can unintentionally increase the disparity between those who have access to digital resources and those who do not, which could exacerbate existing disparities among urban populations. It will take careful thought and inclusive policies that put equity and openness first to address these issues. Involving local communities in the planning process can help ensure that everyone, not just a chosen few, benefits from technological advancements. An all-encompassing strategy that combines technology, policy, and community involvement is necessary for a sustainable and affordable Smart City framework.

This strategy will address urban issues like resource depletion, traffic congestion, and social inequality while fostering environmental sustainability and economic growth. Participation from citizens, data-driven decision-making, and an emphasis on long-term advantages over immediate gains should all be given top priority in this framework. The decision flowchart, as shown in Figure 4.3, illustrates a summary of the expectations for the Smart City of tomorrow.

Figure 4.3: The flowcharts for the sustainable and cost-effective Smart City

Cities may use technology to solve urban problems, improve the lives of their citizens, and build a more resilient and affluent future by embracing a comprehensive and sustainable framework.

All parties involved in this framework, the public and corporate sectors as well as neighborhood associations, must pledge to cooperate in order to realize the common goal of creating a smart and sustainable city.

Sustainable and Cost-Effective Smart City

To achieve a sustainable future, it is essential to address environmental concerns, promote social equity, and ensure economic viability. This multifaceted approach requires a careful balance between ecological preservation, social justice, and economic sustainability. By optimizing resource allocation, we can make informed decisions that protect natural ecosystems while also meeting the needs of diverse communities.

This includes investing in renewable energy sources and sustainable farming practices that not only reduce carbon footprints but also provide fair wages and working conditions for those involved. Efficient use of technology plays a critical role in this strategy. Precision farming and smart grids are two examples of innovative solutions that can significantly lower waste and resource usage.

Long-term success also depends on initiatives being financially feasible and accessible to all parties involved. Engaging with local communities to learn about their needs and ensuring that financial gains are shared fairly are two ways to achieve this. A sustainable and inclusive future can be achieved by combining these components to create robust systems that benefit both people and the environment.

Sustainable Smart City

In order to meet the needs of present residents without sacrificing the ability of future generations to meet their own needs, smart sustainable cities combine sustainable practices with smart technologies. They also adhere to environmental regulations and make use of information and communication technology (ICT) to support their efforts. Through better public services, reduced environmental impact, and effective resource management, these communities prioritize improving the quality of life for their citizens. Smart sustainable cities seek to build resilient urban settings that promote social justice, economic prosperity, and environmental stewardship by utilizing data and creative solutions.

In addition to addressing current urban issues, this all-encompassing strategy encourages long-term sustainability by involving citizens in the decision-making process. Communities may actively shape their urban landscapes as they grow more informed and connected, which will ultimately result in a more dynamic and inclusive society.

Cost-Effective Smart City

The unprecedented rate of urbanization is driving up demand for public services, infrastructure, and housing. Cities face issues including traffic congestion, overpopulation, and environmental deterioration as they grow. Addressing these challenges and creating livable urban environments requires innovative solutions and sustainable planning strategies. However, development is frequently hampered by limited resources and rising costs. Affordable solutions for Smart Cities tackle these issues by:

- **Optimizing Resource Utilization**: Sustainable urban growth depends on reducing waste in energy, water, and materials. Smart technologies like IoT sensors and data analytics can help cities increase efficiency and lower operating costs while also improving the quality of life for citizens by providing them with real-time information on traffic, public transport, and energy consumption, which will ultimately create a more resilient and connected urban community.

- **Improving Efficiency**: Using technology to raise the standard of public services can enhance efficiency. In order to meet the increasing needs of urban populations, public services must be improved by utilizing technology. Municipalities can provide services more efficiently and ensure that citizens have prompt access to information and assistance by optimizing procedures and incorporating digital platforms.
- **Lowering Long-Term Costs**: Investing in long-lasting, low-maintenance infrastructure can lower long-term costs. In addition to reducing ongoing repair costs, investing in long-lasting, low-maintenance infrastructure promotes sustainability by minimizing waste. By using this strategic approach, communities may allocate resources more effectively, improving the lives of their citizens and encouraging environmental stewardship in the process.

One excellent illustration of how nations can carry out extensive Smart City projects while maintaining cost-effectiveness is India's Smart Cities Mission. Prime Minister Shri Narendra Modi launched the Smart Cities Mission on June 25, 2015, with the goal of improving the standard of living in 100 selected cities by providing effective services, strong infrastructure, and a sustainable environment. There are two main methods used to carry out the Smart City Mission. First, as part of the Smart Cities Mission, 100 cities have each selected a specific area for focused interventions, and these cities are being developed using the Area-Based Development (ABD) approach.

As replicable models for other parts of the city, these ABD regions were selected through citizen participation. Second, each city has implemented technology-driven solutions known as Pan-City Projects. Key aspects of this approach include developing a Special Purpose Vehicle (SPV) structure for program implementation, leveraging multiple funding sources for projects, promoting competitive federalism, and engaging citizens.

4.6 Case Studies of Sustainable and Cost-Effective Smart City Initiatives

Smart City projects that employ technology to maximize resource utilization, boost productivity, and improve citizens' quality of life can be both economical and sustainable. Intelligent transportation networks, smart lighting systems, and waste management programs are a few examples. In order to make better decisions and enhance municipal operations, these projects frequently make use of IoT sensors, data analytics, and artificial intelligence.

Smart Energy Management in Urban Infrastructure

Building energy management systems are made up of both monitoring and control components. The building itself, as well as its loads, appliances, and actuators, are closely linked with cyber components such as smart sensors, digital electronic hardware, and software.

Modern buildings are equipped with powerful monitoring, computing, and control features. The various parts of the building's operations are automatically controlled, and monitoring is carried out by analyzing sensor data. Since these buildings include built-in intelligence and automation, they are also known as smart homes or smart buildings. The "sense, analyze, and respond" steps of the SMART strategy are implemented in modern buildings.

Interestingly, designing and maintaining urban systems to use less energy while providing the same level of service, reducing costs, and minimizing environmental impact is known as energy efficiency in urban infrastructure. To reduce energy consumption and encourage sustainable habits, this concept covers a variety of topics, such as infrastructure development, transportation, and building design (Umoh et al., 2024). Cities may become more resilient and adaptable to the changing environment by integrating smart technologies and renewable energy sources. To ensure that residents support sustainable urban policies and actively participate in energy-saving initiatives, it is essential to foster community awareness and engagement. Energy-efficient solutions can also be adopted more widely if laws and policies that support energy efficiency in urban planning and development are put into place.

Examples of these include building codes, energy efficiency standards, and incentives for sustainable practices. One of the most effective ways to reduce costs and enhance efficiency is by implementing smart grids, energy-efficient buildings, and renewable energy sources to lower energy consumption and emissions. Examples include:

- **Adaptive Lighting**: Installing streetlights that adjust brightness based on pedestrian activity and ambient light can save significant energy while improving safety. Switching to LED streetlights further reduces electricity use, lowers maintenance needs, and cuts carbon emissions, thanks to their longer lifespan and higher efficiency compared to conventional lighting.
- **Solar Rooftop Initiatives**: Encouraging residential and commercial buildings to adopt solar panels helps generate clean energy locally, reducing reliance on fossil fuels.
- **Smart Energy Grids**: Integrating smart grids allows for dynamic energy management. When combined with adaptive lighting and other smart systems, energy consumption can be optimized in real time, further enhancing sustainability.

Lowering energy use and emissions through the use of renewable energy sources, smart grids, and energy-efficient structures is a key principle of sustainable and cost-effective smart cities, as illustrated in Figure 4. This strategy encourages innovation in energy management techniques while simultaneously improving sustainability. Through the integration of cutting-edge technologies, communities can maximize energy efficiency and foster environmental responsibility. In addition to generating new employment opportunities in developing green industries, this cooperative effort has the potential to result in significant cost reductions for both businesses and consumers.

Future generations will ultimately benefit from a healthier planet as a result of the shift to a more sustainable energy landscape. Therefore, urban infrastructure that prioritizes energy efficiency can help cities build more resilient, sustainable, and habitable environments for their citizens while also fostering a more sustainable future.

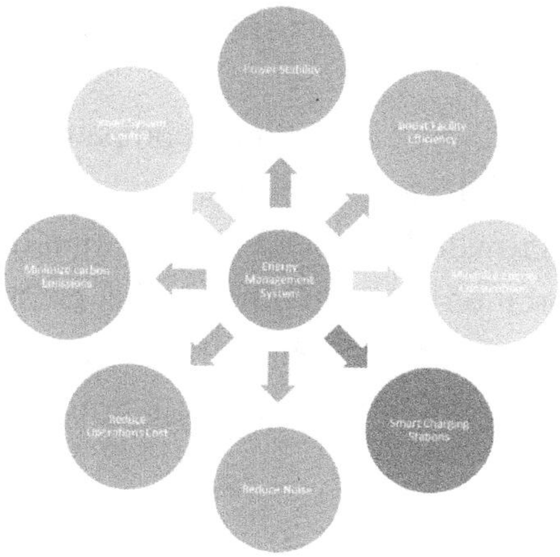

Figure 4.4: Smart City Energy Management System Using Renewable Energy - *Source: Author (2025)*

4.6.1 Case Study 1: Affordable Sustainable Transportation Systems in Lagos

Lagos, one of Africa's fastest-growing megacities, presents

a compelling example of the challenges and opportunities associated with building affordable and sustainable transportation systems. With rapid urbanization, increasing population density, and heavy reliance on road transport, the city faces persistent issues related to congestion, pollution, and infrastructure limitations. This case study explores how Lagos is adopting public transit solutions, smart traffic management, and integrated mobility strategies to improve urban transportation while keeping it accessible, cost-effective, and environmentally sustainable.

Affordable Sustainable Transportation Systems

Encouraging the use of electric cars, public transportation, and bicycle infrastructure in order to lessen pollution and traffic congestion, promote active lifestyles, and reduce dependency on fossil fuels results in healthier communities (Chen and Ma, 2024). Cities can enhance the quality of life for their citizens and contribute to a cleaner environment by investing in sustainable transportation solutions.

Additionally, by attracting companies and generating employment in green technologies and services, these initiatives can boost economic growth. In the end, a dedication to environmentally friendly transportation not only helps the environment but also strengthens community ties. Smart City transportation systems need to serve a large population while remaining reasonably priced. For example:

- Creating affordable Bus Rapid Transit Systems (BRTS) for urban mobility is one way to solve the problem.
- Encouraging the usage of electric buses and rickshaws to lessen reliance on gasoline.
- Putting in place AI-driven traffic control technologies to reduce fuel waste brought on by traffic.

Because of its particular challenges and continuous efforts to enhance urban mobility, Lagos, a commercial city in Nigeria, offers an interesting case study for transportation management. The city is a classic illustration of the challenges faced by rapidly urbanizing megacities due to its severe traffic congestion, poor infrastructure, and dependency on road transportation (Bello and Usifo, 2009). The experience of Lagos emphasizes the necessity of all-encompassing, integrated solutions, such as enhanced infrastructure, intelligent traffic control, and public transportation. The main issues include:

- **Traffic Congestion**: A well-known problem in Lagos, where commuters frequently spend a lot of time stuck in traffic.
- **Inadequate Infrastructure**: Poor road conditions, a dearth of parking spots, insufficient drainage, and other necessary infrastructure.
- **Dependency on Road Transport**: The primary means of transportation in the absence of a strong rail or sea-based mass transit system is the road, which exacerbates traffic.
- **Lack of Integration**: Inefficient and delayed transportation caused by the poor integration of various modes of transport, such as buses, ferries, and taxis.
- **Driver Behavior**: Congestion and accidents are partly caused by bad driving practices and disregard for traffic laws.

The Lagos State Traffic Management Authority (LASTMA) is the main organization in charge of traffic control and enforcement in Lagos. Its efforts are aimed at improving traffic management. In an effort to increase the effectiveness of public transportation, Lagos introduced the first Bus Rapid Transit (BRT) system in Africa

(Kaenzig et al., 2010). Infrastructure development in Lagos is focused on upgrading roads, including bridge and drainage improvements, the construction of new roads, and ongoing maintenance. At the same time, LASTMA is working to educate the public, promoting safer driving habits and raising awareness of traffic laws. Examples of these efforts include:

- **Lagos Rail Mass Transport (LRMT)**: To further reduce traffic and offer an alternative to driving, a rail-based mass transport system is being implemented.
- **Smart Traffic Management**: To enhance traffic flow and safety, the city is exploring the application of intelligent transport systems (ITS).

Key Takeaways and Possible Remedies

- **Combining Different Modes of Transportation**: It is essential to have a smooth and integrated public transportation system that links various forms of transportation.
- **Infrastructure Investment**: It is crucial to make ongoing investments in road upkeep, new road development, and other infrastructure initiatives.
- **Smart Traffic Management Systems**: By putting ITS into place, drivers can get real-time information, monitor road conditions, and optimize traffic flow.
- **Behavioral Change**: Public awareness programs are required to encourage safer driving practices and adherence to traffic laws.
- **Resolving Corruption and Inefficiency**: For efficient transportation management, problems like corruption and a lack of agency cooperation must be resolved.

In conclusion, the difficulties Lagos faces in managing its transportation system and the continuous attempts to resolve them provide important insights into the intricacies of urban mobility in rapidly expanding megacities. The experience of the city emphasizes the necessity of an all-encompassing, integrated strategy that includes smart technologies, public transportation, infrastructure development, and behavioral modification. Equally important is learning from best practices, as valuable insights can be gained by studying effective transport management techniques in other cities, including Dubai.

4.6.2 Case Study 2: Singapore's Smart Waste Management

Singapore is widely recognized as a global leader in urban cleanliness and environmental sustainability, making it an ideal example for examining the role of smart technologies in waste management. Through the strategic adoption of Internet of Things (IoT) solutions, Singapore has transformed traditional waste collection practices into a highly efficient, data-driven system that emphasizes real-time monitoring, optimized resource use, and reduced environmental impact. This case study highlights how the city-state leverages technology to address the growing challenges of urban waste, setting a practical benchmark for smart and sustainable waste management in modern cities.

Smart Waste Management

The effective management of waste has emerged as a crucial concern for local authorities in an increasingly urbanized world. Conventional waste management systems frequently find it difficult to keep up with the expanding urban population, resulting in inefficiencies, increased environmental impact, and rising expenses. However, the use of Internet of Things (IoT) technology is transforming urban waste management and providing a pathway to clean, efficient, and sustainable urban environments. Effective urban waste management is essential for both public health and the overall standard of living in cities. Local governments can optimize collection routes, monitor waste levels in real time, and ultimately reduce costs while minimizing their environmental impact by deploying IoT-enabled sensors and smart waste bins.

For the purpose of enhancing waste processing and achieving greater cost-effectiveness, sustainability, and efficiency, smart waste management makes use of data and technology. This approach includes monitoring and managing waste from sorting to disposal, collection, transportation, and recycling through the use of sensors, GPS, analytics, and automation. Additionally, to further increase sustainability and expedite waste collection, smart waste management leverages technologies such as the Internet of Things (IoT).

Implementing recycling initiatives and intelligent waste collection systems helps reduce waste and enhance resource utilization. These programs support community awareness and environmental sustainability in addition to reducing landfill overflow. Cities can ensure that resources are used effectively and efficiently by optimizing collection routes using technology.

The impact of these programs can be further amplified by involving residents in educational efforts that promote recycling and proper waste management. In the end, everyone can benefit from a cleaner, more sustainable future if corporations, cities, and individuals work together. One important but frequently overlooked component of urban development is waste management. Figure 4.5 depicts the usage of smart waste bins by cities to achieve cost-effective waste management, through:

- Installation of smart bins that alert collection teams when they are full.
- Establishing plants that convert waste into energy to produce electricity from solid waste.

- Promoting source-side categorization by means of education initiatives.

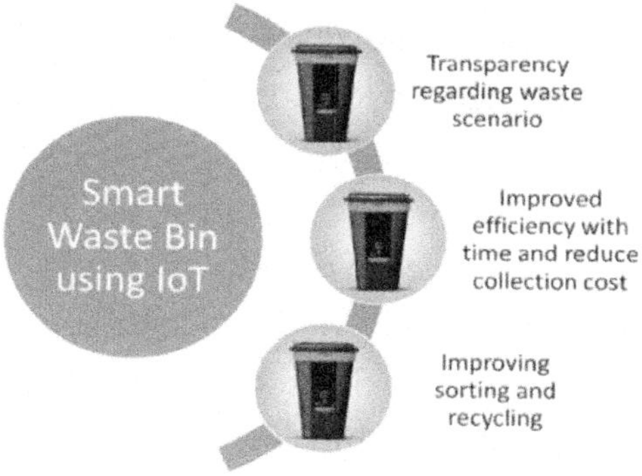

Figure 4.5: Smart Waste Management Using IoT

Singapore has set a benchmark for the use of IoT in enhancing waste management (Ferro-Escobar, 2022). The Internet of Things is essential to preserving Singapore's reputation for being immaculately clean. The city-state uses a cutting-edge waste collection system with sensors that can detect odors and monitor bin fill levels. Real-time tracking is made possible by these sensors, which is crucial in a densely populated city like Singapore.

The city has significantly reduced operating costs and environmental impact by minimizing unnecessary trips and only emptying dumpsters when needed. In addition to improving waste management efficiency, this innovative strategy helps create a more sustainable urban environment. Singapore's dedication to cleanliness and environmental stewardship sets a global benchmark as it continues to adopt smart technologies.

4.6.3 Case Study 3: Smart Water Management Systems in Urban Cities

With growing urban populations and increasing pressure on limited water resources, smart water management has become a critical priority for modern cities. Traditional water systems often struggle with inefficiencies, leakages, and poor demand forecasting, leading to significant resource losses and higher operational costs. This case study examines how the integration of digital technologies, sensors, and real-time data analytics can transform urban water management into a more efficient, resilient, and sustainable system.

Smart Water Management Systems

Leveraging data and technology, intelligent water management in cities maximizes the use of water resources while boosting sustainability and efficiency. Sensors, Internet of Things devices, and real-time data analysis are used to track water consumption, detect leaks, and improve water distribution.

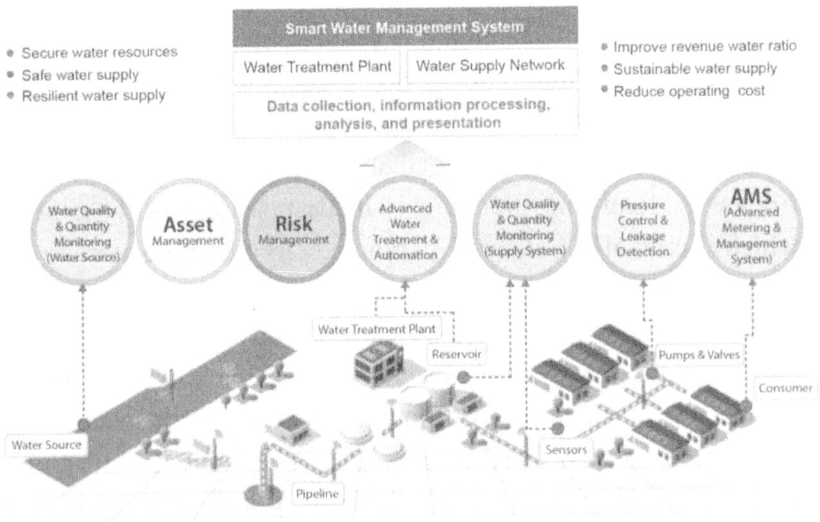

Figure 4.6: You Kwangtae, CEO, UnU Civil & Environmental Engineering, Republic of Korea

Smart water management systems can also address water quality issues, wastewater treatment, and water demand forecasting. Municipalities can save a significant amount of money and reduce their overall environmental impact by implementing smart water sensors, leak detection systems, and water conservation techniques to improve water efficiency. Cities can further strengthen their efforts toward sustainable water management by encouraging community involvement and raising awareness of water-saving practices.

In addition to saving money and enhancing sustainability, smart water management systems can provide a more robust and efficient water delivery system. Digital meters and sensors, Geographic Information Systems (GIS), and Supervisory Control and Data Acquisition (SCADA) systems are examples of high-tech solutions for the water sector. Water utilities can install sensors to monitor operations and collect data in real time, as well as create a digital representation of their water distribution network.

Conventional water and wastewater systems can be upgraded into instrumented, networked, and intelligent systems through the use of smart technologies:

- **Instrumented:** Equipped with sensors and devices that can measure, record, and detect relevant data.
- **Interconnected:** Able to communicate and interact with system managers and operators for coordinated monitoring and control.
- **Intelligent:** Capable of analyzing situations, responding quickly, and optimizing problem-solving and maintenance processes.

The explanation of Figure 4.6 is based on the proceedings of an Asian Development Bank (ADB) workshop on the water sector held in Tashkent, Uzbekistan, which highlighted smart systems and focused on remote water network monitoring using smart meters and other technologies.

4.6.4 Case Study 4: Smart and Affordable Housing in Urban Development

As cities continue to expand and urban populations grow, the demand for affordable and sustainable housing has become a central challenge for urban planners. Smart and affordable housing integrates technology, inclusive design, and sustainable construction practices to address housing shortages while improving living standards for low and middle-income communities. This case study explores how urban planning strategies and smart city principles can be combined to create equitable, efficient, and resilient housing solutions in rapidly developing urban environments.

Low-Cost Housing and Urban Planning

Affordable housing is essential to urban planning because it meets the housing demands of low- and middle-income residents, particularly in rapidly urbanizing areas. A comprehensive strategy that incorporates needs analysis, inclusive design, stakeholder participation, and sustainable practices is necessary for the successful integration of affordable housing into urban planning.

In rapidly urbanizing regions like Lagos, Nigeria, in West Africa, affordable housing must be incorporated into urban planning to create inclusive and sustainable cities. Low-cost housing can help build a more equitable and resilient urban environment by meeting the housing needs of low-income residents, promoting social cohesion, and supporting economic growth. One of the key elements of Smart Cities is affordable housing. Key strategies include:

- Using prefabricated building techniques to reduce costs and construction time.
- Promoting mixed-use development to maximize land utilization.

- Employing eco-friendly building materials, such as recycled steel and fly ash bricks.

There is little opportunity for ordinary citizens to participate in smart cities due to the influence of profit-driven global tech firms, the alignment of these firms with local government agendas, and a competitive form of "urban entrepreneurialism" (Hollands, 2015). Projects such as Guangzhou's Knowledge Metropolis in China, Songdo in South Korea (a purpose-built, high-tech, environmentally friendly commercial city), Masdar in the UAE, and Singapore's 2015 Intelligent Nation initiative, are primarily designed to attract talent, skilled labor, and knowledge-based industries. However, as Sennett (2012) warns, "the danger now is that this information-rich city may do nothing to help people think for themselves or communicate well with one another."

Successful implementation requires careful planning, collaboration among stakeholders, and addressing potential challenges such as cost and equity. By offering accessible, sustainable housing options and leveraging technology to improve efficiency, smart and affordable housing initiatives in smart cities aim to enhance residents' quality of life. Ultimately, these efforts can help create more vibrant, inclusive, and welcoming urban environments. Regulatory barriers, high land acquisition costs, and construction expenses can all pose challenges to implement affordable housing in smart cities. It is essential to ensure that smart city technologies promote equity and do not inadvertently create or exacerbate social injustices. When used thoughtfully, technology can enhance mobility, reduce energy consumption, optimize resource management, and encourage community engagement.

In this context, affordable housing plays a key role by providing accessible, sustainable living options for a broader segment of the population, supporting the overall goals of smart city initiatives. One effective strategy for creating more livable and sustainable urban environments is the implementation of smart, affordable housing programs in smart cities. By thoughtfully addressing challenges and prioritizing equity and collaboration, these initiatives can contribute to a more inclusive and prosperous urban future.

4.6.5 Case Study 5: Green Areas and Urban Forestry in Smart Cities

As urbanization accelerates and cities face increasing environmental pressures, the integration of green areas and urban forestry has become a vital component of sustainable urban development. Green spaces not only enhance the visual appeal of cities but also play a critical role in improving air quality, reducing urban heat, supporting biodiversity, and promoting public well-being. This case study explores how smart cities are using technology, policy frameworks, and community engagement to protect and expand urban green infrastructure, creating healthier and more resilient urban environments.

Green Areas and Urban Forestry

Incorporating green spaces and urban forestry into city landscapes enhances biodiversity, mitigates the urban heat island effect, and improves air quality. Beyond their aesthetic appeal, these natural elements are vital for creating resilient and sustainable cities. They increase urban areas' capacity to adapt to climate change, promote mental and physical well-being, and improve both air and water quality.

A Smart City can make urban areas more resilient, sustainable, and livable by incorporating green spaces and urban forestry. For sustainable and economically viable smart cities, these features are essential because they provide a range of socio-economic and ecological benefits. They contribute to lower energy consumption, improved water and air quality, increased biodiversity, and enhanced urban cooling. Particularly in resource-constrained settings, smart technologies such as open data and community participation can strengthen the management and sustainability of these areas.

Figure 4.7: Green Area Mapping of Ahmadu Bello University, Main Campus, Zaria – Nigeria; *Source: Aondofa (2021)*

For both faculty and students at every university, a high-quality learning environment is essential, and selecting the appropriate land area for green spaces is one of the most crucial aspects of campus planning and design. At the main campus of Ahmadu Bello University in Samaru, Zaria, Nigeria, Figure 4.7 illustrates a portion of the mapped land-use distribution, focusing on the sustainability of green area networks.

The eco-friendly community in Singapore, widely regarded as a global leader in intelligent and sustainable urban development, is combining green projects with technological advancements to enhance living standards and reduce their negative effects on the environment.

These initiatives, launched in 2014 as part of the Smart Nation initiative, demonstrate Singapore's determination to incorporate cutting-edge technologies. For example, the integration of artificial intelligence (AI) and the Internet of Things (IoT) into public services, governance, and urban planning remains a key focus. This initiative aims to create a more sustainable, integrated, and effective urban environment (Khern, 2019; Caragliu and Del Bo, 2018; Allam and Dunny, 2019). Singapore has made significant strides in smart urban living from 2015 to 2023. This progress is exemplified by initiatives such as Tengah Eco-Town and the Punggol Digital District, which combine digital solutions and sustainable infrastructure to enhance the daily lives of residents. The government has also actively partnered with tech giants like Google and Alibaba to expand its digital capabilities and foster a collaborative approach to urban development (Du et al., 2015). The new Tengah Eco-Town, Smart City projects, and an emphasis on sustainability through technology and green spaces are all important components of Singapore's green Smart City strategy.

Figure 4.8: Singapore Eco-Smart City; *Courtesy: Housing & Development Board (Holland, 2021)*

In addition to enhancing urban life, these initiatives support biodiversity and foster citizen involvement, positioning Singapore as a role model for other cities seeking to balance environmental stewardship with urbanization. As a result, residents can enjoy a high quality of life while contributing to a more environmentally sustainable future. Initiatives such as massive park networks, vertical gardens, and energy-efficient buildings, all of which are aimed at promoting a harmonious coexistence of nature and urban life, are examples of this commitment to sustainability. In addition to addressing the challenges of urbanization, Singapore provides a positive example for cities worldwide by integrating eco-friendly practices with smart technologies. This dedication has inspired other countries to follow suit, demonstrating that environmental conservation and urban growth can coexist. With cities across the globe grappling with the impacts of population growth and climate change, Singapore's model serves as a shining example of innovation and adaptability, fostering a global movement toward sustainable urban living.

Large green areas, such as "forest corridors," which link nearby nature reserves and urban areas, are another feature of Tengah's design, as shown in Figure 4.8. In addition to improving biodiversity, these green areas also reduce urban heat islands, improve air quality, and provide locals with access to recreational opportunities. Walking, bicycling, and public transportation are also given priority in the town's car-free zones, which reduces traffic emissions and encourages healthier living.

In addition to promoting a sense of community, this all-encompassing approach to urban planning encourages citizens to adopt sustainable lifestyles. As Tengah continues to develop, it aims to set an example for other cities that prioritize both quality of life and environmental integrity. In addition to tackling urgent environmental challenges by improving air quality and mitigating urban heat islands, Tengah's innovative approach to urban design enhances the community's well-being by providing a wide range of recreational opportunities.

The creation of car-free zones encourages walking, cycling, and the use of public transportation, promoting healthier lifestyles for residents. By emphasizing these priorities, Tengah establishes a benchmark for sustainable living that other communities can aspire to achieve. This comprehensive vision demonstrates how thoughtful urban planning can harmonize environmental stewardship with community well-being, paving the way for future developments that put people and the planet first.

Example from regional concepts

Several regions offer valuable insights into economical and sustainable Smart City frameworks. Cities like Amsterdam focus on circular economies, while Vienna emphasizes electrification. In Europe, Copenhagen and Oslo are leaders in zero-emission initiatives and urban sustainability. Meanwhile, Latin American cities such as Medellín and Santiago leverage technology to transform their communities while prioritizing resilience, safety, and social diversity. Smaller cities and eco-districts also provide lessons on the power of community involvement and the effectiveness of targeted, locally driven solutions.

In Europe, digital technology integration and multi-stakeholder partnerships are prioritized, particularly through programs such as the European Innovation Partnership for Smart Cities and Communities. While context-specific approaches are essential, Latin America highlights the importance of locally applicable models and holistic urban planning. Even smaller communities, which are often overlooked, can benefit by focusing on effective waste management, accessible public transportation, and active citizen participation (Yang and Zhen, 2024).

European Innovation Partnership (EIP) on Smart Cities and Communities

This program fosters collaboration among cities, businesses, academic institutions, and residents to accelerate the development of smart cities. By pooling resources and expertise, it aims to implement innovative solutions that enhance sustainability and urban living. Through the effective use of data and technology, cities can strengthen community connectivity, reduce environmental impact, and improve the quality and efficiency of public services.

Smart City Lighthouse Projects

Horizon 2020, a European research and innovation program, funds projects aimed at demonstrating scalable and impactful solutions in areas such as energy, mobility, and information and communication technology (ICT). These projects often involve collaboration between cities, industry, and citizens, emphasizing participatory and multi-stakeholder approaches. The initiative particularly supports projects that focus on advanced ICT infrastructure, sustainable mobility, energy efficiency, and the integration of diverse technologies to create more livable and sustainable urban environments.

A key component of the program is its "Smart Cities and Communities Lighthouse" initiatives, which showcase large-scale, extremely smart city solutions that serve as models for other communities. Ultimately, Horizon 2020 seeks to foster the development and deployment of innovative technologies that enhance economic growth, mobility, quality of life, and energy efficiency across European cities.

Vienna's Smart City Framework Strategy

To build a smart, livable, and environmentally sustainable city, Vienna has developed a comprehensive urban development plan that addresses energy, mobility, and citizen engagement. This plan, part of the Smart City Wien Framework Strategy, aims to enhance quality of life, promote the adoption of renewable energy, and reduce greenhouse gas emissions.

The strategy outlines goals for creating a city that integrates and prioritizes energy systems, mobility solutions, building standards, and infrastructure, ensuring a cohesive approach to sustainable urban development. According to the European Commission, Vienna's strategy exemplifies a comprehensive vision for sustainable urban development, integrating social, economic, and environmental considerations to create a resilient, inclusive, and livable city.

Multi-Stakeholder Partnership

To develop effective and sustainable solutions to urban challenges, the European Union (EU) strongly promotes multi-stakeholder partnerships in Smart City initiatives, emphasizing collaboration among local governments, businesses, citizens, and other relevant actors. By bringing together diverse perspectives and expertise, this approach enhances ownership and commitment, improves transparency and accountability, better aligns projects with community needs, and fosters innovation and sustainable development.

Through this emphasis on participatory partnerships, the EU is advancing a more inclusive approach to urban development, recognizing that the active engagement of all relevant stakeholders is critical to the success of Smart City initiatives.

Digital Technology Integration

Europe recognizes the significant potential of digital technologies to enhance multiple aspects of urban life, particularly in public services, energy management, and transportation. The European Innovation Partnership on Smart Cities and Communities (EIP-SCC) is a key initiative that advances this vision by connecting diverse stakeholders to promote innovation and transformation in urban environments.

A central component of this effort is energy management systems, which focus on optimizing energy distribution, storage, and consumption. These systems aim to achieve efficient resource use, support climate protection, and reduce economic costs.

Latin American Outlook

Rather than adopting smart city models from other countries, Latin American cities are encouraged to develop approaches tailored to their specific social, economic, and environmental contexts. This localized strategy promotes more effective and sustainable urban development by recognizing the unique opportunities and constraints within each city.

Latin American cities face particular challenges, including environmental vulnerabilities, economic disparities, and social inequalities (Prateeppornnarong, 2025). As a result, "one-size-fits-all" solutions from other regions may be unsuitable or ineffective in addressing these complex, context-specific issues. For example, Colombia's Medellín has implemented Smart City initiatives that prioritize green spaces and public transportation while adapting solutions to the local environment (Angelidou, 2017). Similarly, Barranquilla's "Todos al Parque" program addresses community-specific needs by improving urban green spaces and enhancing residents' well-being.

Latin American cities continue to face challenges such as limited capacity, uneven technological adoption, and insufficient infrastructure. Addressing these issues requires targeted strategies and strong partnerships. By focusing on context-specific approaches, communities in Latin America can leverage smart city technologies to tackle their unique challenges, fostering more sustainable, equitable, and prosperous urban environments.

4.7 Reflections on Smaller City

Public Transport Accessibility

By leveraging intelligent mobility technologies such as demand-responsive transport (DRT) and integrated service platforms, smaller cities can prioritize accessible and efficient public transit systems. These strategies enhance sustainability, improve mobility, and address the unique transportation challenges faced by smaller urban areas.

Integrating multiple modes of transport such as buses, ridesharing, and bike-sharing programs into a single platform can be especially beneficial for these communities. Users may be encouraged to choose public transportation over private vehicles due to the smooth and convenient experience provided by intelligent mobility systems. Smaller cities are particularly well-suited for demand-responsive transport (DRT) systems, which dynamically adjust schedules and routes based on real-time passenger demand. In areas with lower population densities or underutilized roads, DRT can optimize routes, reduce waiting times, and improve overall service efficiency, making public transit a more attractive and sustainable option.

To enhance the efficiency and user experience of public transportation, smaller cities can leverage smart technologies such as mobile ticketing, real-time data analytics, and intelligent traffic management systems. These tools enable route optimization, demand forecasting, and the provision of accurate, up-to-date information to riders, making transit systems more reliable, convenient, and attractive for citizens.

Smart Waste Management

Implementing smart waste management systems can promote recycling, reduce costs, and optimize collection routes.

Emphasis On Citizen Participation

By leveraging diverse platforms and engagement strategies, smaller cities can foster a sense of community and encourage active citizen participation in smart city initiatives. In conclusion, there is broad consensus that Smart City initiatives must integrate sustainability and resilience strategies to ensure that cities can adapt to evolving conditions and emerging challenges. By harnessing data from multiple sources, urban authorities can enhance public services, optimize resource allocation, and improve the overall quality of life for their residents.

The development of Smart Cities requires collaboration across multiple sectors, including government, private companies, academic institutions, and local residents. Smart city initiatives should comprehensively address all aspects of urban life, including the economy, environment, governance, society, mobility, and overall quality of living.

4.8 Benefits of the Framework for Sustainable and Cost-Effective Smart City

A framework for sustainable and affordable Smart Cities offers numerous benefits, including improved public services, reduced traffic congestion, lower energy consumption, and an overall higher quality of life. Additionally, it fosters innovation, encourages citizen participation, and stimulates economic growth. By leveraging technologies such as IoT sensors and data analytics, smart cities can enhance multiple aspects of urban life, creating more efficient, comfortable, and sustainable living environments for their residents.

By addressing the needs of growing urban populations, smart city frameworks leverage ICT to enhance urban environments, improve quality of life by 10% to 30%, and promote sustainable urbanization (Khan and Labonté, 2021). Cost-effective and sustainable smart cities provide a wide range of benefits, including reduced resource consumption and environmental impact, increased inclusivity, long-term savings through lower operating and maintenance costs, and robust economic growth. These benefits can be broadly categorized as follows:

How Smart Cities Improve Quality of Life?

- **Response City Services** - Smart city technologies can make public services such as waste management, transportation, and safety more efficient and responsive to citizens' needs.

- **Decreased Traffic Congestion**: Intelligent traffic management systems optimize traffic flow, helping to shorten commuting times and ease congestion.

- **Highways and streets that are safer**: Real-time data analysis, smart lighting, and surveillance systems enhance public safety and help reduce crime.
- **Improved Citizen Engagement**: Online participatory tools allow residents and local authorities to communicate and collaborate more effectively, fostering stronger community involvement.

Sustainability and Environmental Benefits

- **Lower Energy Consumption**: Smart energy networks and building management systems enhance energy efficiency and minimize energy waste.
- **Reduced Carbon Footprint**: Smart Cities help cut carbon emissions by promoting sustainable transportation options and reducing overall energy consumption.
- **Better Waste Management**: Intelligent waste management systems encourage recycling, reduce waste volumes, and optimize collection routes, leading to a healthier environment, more efficient resource use, and lower pollution levels.

Economic Benefits

- **Quicker Economic Growth**: Smart Cities can attract investments, foster the development of new economic sectors, and create employment opportunities, thereby stimulating overall economic activity.
- **Enhanced Efficiency**: Leveraging data-driven decision-making and automation helps streamline processes across various sectors, improving operational efficiency.
- **Cost Savings**: Improved public services, optimized resource use, and reduced energy consumption can generate substantial cost savings for both the city and its residents.

Further Benefits

- **Enhanced Innovation**: Smart Cities provide a platform for testing and implementing novel technologies, fostering creativity and technological advancement.

- **Increased Citizen Awareness and Participation**: By giving access to real-time information and data, Smart Cities empower individuals to make informed decisions and engage actively in civic life. Equal access to opportunities and services also promotes social inclusivity.

- **Improved Governance**: Smart City frameworks support more transparent, accountable, and responsive governance, strengthening trust between citizens and authorities.

4.9 Challenges and Solutions in Implementing Sustainable and Cost-Effective Smart City

Creating Smart Cities that are both economical and sustainable requires balancing social and environmental priorities, technological innovation, financial constraints, and potential inequalities. Key challenges include the high cost of infrastructure, concerns over data privacy, potential negative social effects such as gentrification, and the need for robust governance and regulatory frameworks to ensure equitable access to opportunities and technology.

By addressing these challenges and considering these key factors, cities can strive to create sustainable and affordable Smart Cities that benefit all citizens while fostering a more resilient and equitable future. While the concept of an economical and sustainable Smart City is promising, several issues still need to be resolved:

- **Limited Funding**: Implementing Smart City technologies can be expensive and requires careful budgeting and planning. Identifying affordable and sustainable sources of financing is essential for the successful execution of large-scale projects.

- **Technological Barriers**: Ensuring that new solutions are compatible with existing infrastructure is critical. It is also necessary to guarantee data security, protect citizen privacy, and provide the technology and infrastructure needed to support Smart City initiatives.
- **Engaging the Community and Public Acceptance**: Gaining the support and trust of residents is crucial. Smart City projects must benefit all citizens, not just privileged groups, and active community involvement ensures inclusivity, transparency, and long-term success.

To effectively address these challenges, a close collaboration among local communities, private sector actors, and government institutions is essential. By fostering cooperation and the sharing of resources, stakeholders can jointly develop innovative solutions that advance the three foundational concepts of viability, equitability, and bearability. These concepts are often used to define the core dimensions of sustainable development, which emerge at the intersection of the three main pillars of sustainability environmental, economic, and social as shown in Figure 4.1.

This coordinated approach not only addresses current challenges but also lays a strong foundation for long-term resilience and sustainable prosperity. This collaborative effort can foster a stronger sense of community, enhance the delivery of public services, and promote greater economic stability.

Ultimately, by working toward shared objectives, all stakeholders can contribute to building a more prosperous, inclusive, and equitable society for future generations. By prioritizing inclusivity and open communication, these collaborations can adapt to evolving needs and emerging challenges over time. This approach empowers communities, fosters a culture of mutual support, and ensures that development remains meaningful, resilient, and sustainable in the long term.

4.10 The Role of Public-Private Partnerships (PPPs) in Implementing Sustainable and Cost-Effective Smart City

Public–Private Partnerships (PPPs) are collaborative arrangements in which government agencies and private-sector entities jointly finance, develop, and operate public infrastructure and services. These partnerships leverage private-sector expertise, innovation, and capital to deliver projects such as roads, public transportation systems, and healthcare facilities, often transferring a significant share of financial and operational risks to private partners. In the context of smart city development, PPPs play a critical role in enabling the efficient financing, implementation, and management of technology-driven urban initiatives.

Public–Private Partnerships (PPPs) enable cost sharing and reduce pressure on public finances while providing access to private-sector expertise and advanced technologies. Through streamlined and efficient processes, PPPs often accelerate project delivery and improve service quality.

By leveraging private-sector knowledge, innovation, and financial resources while sharing risks with the public sector PPPs play a vital role in the implementation of sustainable, resilient, and economically viable smart city initiatives.

This partnership enhances citizen participation in urban planning and decision-making processes, accelerates the development of Smart Cities, and improves overall project efficiency. By fostering collaboration between public and private sectors, Public–Private Partnerships (PPPs) enable the pooling of financial resources, technical expertise, and innovative capacity, thereby supporting the successful implementation of smart city initiatives. This can be further explained as follows:

- **Cost-Effectiveness and Resource Mobilization**: PPPs give cities access to more resources and knowledge beyond public budgets. By providing funds, technology, and creative solutions, the private sector can lessen the financial strain on taxpayers and possibly reduce project expenses overall.
- **Knowledge and Innovation**: Private businesses frequently possess state-of-the-art equipment and knowledge in fields like data analytics, artificial intelligence, and the Internet of Things, which are essential for the development of smart cities. Cities can leverage these resources and accelerate the adoption of intelligent solutions by collaborating with the private sector.
- **Project Sustainability and Risk Sharing**: PPPs allow the public and private sectors to share duties and risks, which makes large-scale projects easier to manage and more sustainable. Because the public sector is not the only entity assuming the operational and financial risks, more ambitious initiatives can now be undertaken.
- **Improved Citizen Involvement**: By offering opportunities for public input and involvement in project design and execution, PPPs can promote increased citizen involvement in urban planning and decision-making. This results in smart city projects that are more responsive and inclusive.
- **Sustainability and Environmental Impact**: By encouraging environmentally friendly behaviors and technologies, PPPs can support sustainable urban development. PPPs, for instance, can help develop smart grids, sustainable transportation systems, and infrastructure for renewable energy.

Public safety, water management, energy systems, and transportation infrastructure are just a few of the Smart City projects that leverage PPPs. For example, PPPs can fund the construction of smart waste management systems, smart traffic management systems, and infrastructure for electric vehicle charging. These initiatives support efficiency and sustainability in city operations, in addition to improving urban living.

Cities can adopt creative solutions that address urgent issues and promote economic growth by combining resources and knowledge from the public and private sectors. Additionally, these partnerships can draw talent and investment to the region, enhance public services, and improve the quality of life for locals. PPPs will play an ever more important role in shaping urban development in the future, as cities continue to evolve by investing in joint ventures with private businesses to benefit from their resources and expertise.

This strategy can improve project outcomes by combining the efficiency and creativity of the commercial sector with the objectives of the public sector. It also enables joint investment, which lessens the financial strain on government agencies. For instance, the Delhi-Mumbai Industrial Corridor (DMIC) leverages public-private partnerships (PPPs) to develop smart infrastructure across multiple cities.

4.11 The Role of Technology in Implementing Sustainable and Cost-Effective Smart Cities

By facilitating better resource management, boosting public engagement, and improving services, technology is essential to the implementation of sustainable and affordable Smart Cities. Cities can reduce costs and their impact on the environment by optimizing waste management, energy use, and transportation through data-driven solutions and the Internet of Things (IoT). As urban populations continue to rise, technology's role in implementing sustainable and cost-effective Smart Cities becomes increasingly important. Cities can lower operating costs, improve resource management, enhance infrastructure, and reduce their environmental impact by incorporating cutting-edge technology such as IoT, AI, and renewable energy sources. The foundation of economically viable Smart Cities is technology.

Innovative tools, such as AI-driven analytics and IoT-based sensors, are useful for tracking the use of resources and energy in real time, reducing the cost of maintenance by anticipating and preventing infrastructure failures, and minimizing manual intervention while streamlining the delivery of services. A closer look at the main responsibilities is provided below:

Improving the Management of Resources

- **Energy management**: By optimizing energy production, distribution, and consumption, smart grids and sensors can reduce waste and energy costs.
- **Water management**: By monitoring water quality and identifying leaks, smart pipes and meters can reduce water waste and ensure effective delivery.
- **Waste Management**: By optimizing waste collection, smart waste containers and collection routes can reduce emissions and fuel consumption.

Enhancing Public Services

- **Transportation**: By optimizing public transit routes and reducing traffic congestion, intelligent transportation systems can save both time and money.
- **Healthcare**: Access to care can be improved, and costs can be reduced with telemedicine and other tech-enabled healthcare solutions.

- **Education**: Digital technology, such as online learning platforms, can reduce the cost of education while also enhancing educational opportunities.

Improving Public Involvement

- **Participation of Citizens**: Digital platforms enable real-time input and involvement in decision-making processes.
- **Community Communication**: By giving citizens quicker access to information and updates, mobile applications and other digital platforms can improve communication between the government and the public.
- **Public Safety**: By lowering crime rates and improving public safety, CCTV systems and other surveillance technology can be effective.

Facilitating the Development of Sustainable Cities

- **Green technologies**: Smart Cities can reduce their dependency on fossil fuels and promote a cleaner environment by integrating renewable energy sources, such as wind and solar power.
- **Data-Driven Decision Making**: Big data and analytics can offer valuable insights into policymaking and urban planning, facilitating better-informed and sustainable choices.
- **Resilient Infrastructure**: Smart Cities can integrate technologies that enhance their ability to adapt to environmental challenges, such as climate change.

In conclusion, technology is a potent instrument for building economically viable and sustainable Smart Cities by facilitating better resource management, enhancing public services, encouraging citizen participation, and supporting sustainable urban growth. Platforms for Smart Cities, as seen in Bengaluru, demonstrate how technology can increase affordability and efficiency.

4.12 A Proposed Model Smart City Framework for a Developing Economy – A Case of Lokoja City

Lokoja is a city located at the confluence of the Niger and Benue rivers. It serves as the capital of Kogi State in north-central Nigeria, West Africa. It is popularly known as "The Confluence City" and, with its varied, multiethnic population, holds great historical and geographical value. The choice of Lokoja as a case study is unique because of its historical significance, and the city is endowed with a blend of aquatic, terrestrial, and highland features. The major features are:

4.12.1 The Confluence and its Geography

The confluence of the Niger and Benue Rivers defines Lokoja's uniqueness. The clearer, greenish Benue flows southwards beside the muddy, silt-heavy Niger, allowing visitors to witness a dramatic contrast.

Figure 4.9: The sight of Confluence of Niger and Benue rivers in Lokoja - *Source: adapted and modified from Wikipedia*

This unusual confluence provides an important resource for nearby communities, in addition to creating a scenic environment. Because of its abundance of biodiversity, the area attracts tourists interested in the region's cultural heritage as well as those fascinated by its wildlife. Figure 4.9 presents the exact meeting point of the Niger and Benue Rivers.

To increase trade between the Sahelian states of the Republics of Niger and Mali, the Niger River could potentially be developed into an international sea route or canal. Through Lokoja, where it connects to the vast Niger Delta, which joins the Atlantic Ocean and the Gulf of Guinea, the Niger River can serve as an inland waterway for landlocked Sahelian nations such as Mali and Niger. Ships can reach the global shipping network from the Niger Delta; however, to connect to other continents, such as South America, they must cross the Atlantic Ocean rather than using a direct river route.

The administration and development of the river system are complicated by the fact that nine countries share the Niger Basin. The millions of people who depend on the river also rely on its environmental health. Therefore, promoting sustainable growth in this crucial region requires protecting its shipping potential, ecotourism, and ecological integrity. The urban area around the confluence could be transformed into an attractive tourism destination.

4.12.2 Mount Patti

The city is situated on a hill approximately 458 meters tall, which offers a broad perspective over the city below and the river confluence. Tourists frequently flock to the summit to witness stunning sunsets, where the pink and orange tones blend harmoniously with the surrounding scenery. The sounds of music and laughter fill the lively streets below, evoking a charming ambiance that perfectly embodies the spirit of the city.

The city's sparkling lights reflect off the river as night falls, adding to the magical atmosphere. The vibrant local markets offer a wide range of products and cuisine that highlight the region's diverse traditions and culture. The approximate land area of Mount Patti is 15 square kilometers, which could be developed using the business model of Genting Highlands (about 1,800 meters high) in Malaysia.

Figure 4.10: Lord Lugard ancient resting room at Mount Patti, Lokoja - *Source: adapted and modified from Wikipedia*

Figure 4.11: View of Cable Cars of Genting Highlands, Malaysia. *Source: adapted and modified from Wikipedia*

Figure 10 depicts the upper surface of Mount Patti, showing the image of the ancient resting house of Lord Lugard. Figure 4.11 shows the Genting Highlands Resort, designed with an ideal location for shoppers seeking high-end items at competitive prices. Over 150 designers and premium brands are featured at the outlet, including Burberry, Coach, Michael Kors, Polo Ralph Lauren, and many more. The outlet is designed to mimic a small village, with shops and eateries scattered across a vast area. Visitors can stroll along the walkways, enjoying the cool mountain air while browsing the latest fashion trends.

4.12.3 The Central Location of the City

Nigeria's northern and southern regions are connected by its central location. Lokoja is therefore a significant hub for trade and transit. The city's economy has grown considerably as a result, attracting both tourists and businesses. Its appeal as a key center for trade and connectivity is further enhanced by its thriving markets and rich cultural heritage. Visitors from a variety of backgrounds are drawn to Lokoja because of its distinctive atmosphere, which is created by the blending of modernity and tradition.

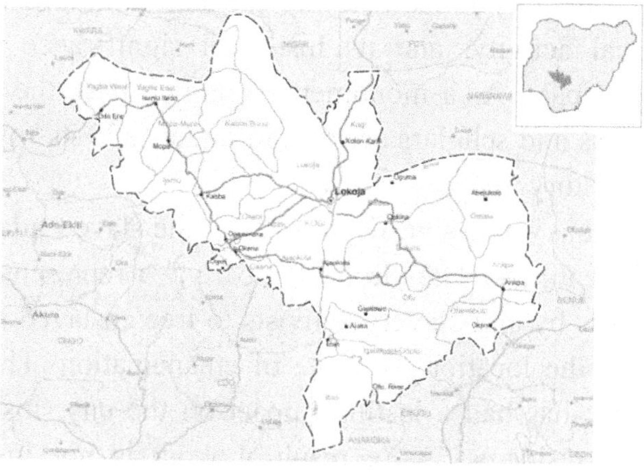

Figure 4.12: Map of Lokoja, Kogi State of Nigeria

Source: OCHA

Its role as a major player in Nigeria's economy is also expected to be strengthened by continued infrastructure investment. Figure 4.12 indicates the position of Lokoja in the maps of Kogi State and Nigeria.

4.12.4 Historical Significance of Lokoja

William Baikie, a British explorer, founded Lokoja as a trading center in 1857 (Wilson, 2020; Ali and Nomishan, 2024). An important advantage for trade and commerce was its location at the confluence. In addition to facilitating trade, this strategic position attracted diverse groups and cultures to the area. Over time, Lokoja developed into a thriving center that was crucial to Nigeria's socioeconomic growth. Sir Frederick Lugard, the High Commissioner at the time, used Lokoja as the administrative center for the Protectorate of Northern Nigeria from 1900 to 1901 (Apata, 1990).

Lagos became the first capital of the unified Colony and Protectorate of Nigeria in 1914 after Lugard officially merged the Northern and Southern Protectorates. Prior to its relocation to Abuja in 1991, Lagos served as the capital. Due to the proximity of Abuja to Lokoja, the city remains a hub for both cultural exchange and commercial activity, and its historical significance has left an enduring impact. As a monument to its rich past, the city attracts both tourists and scholars seeking to understand its pivotal role in Nigerian history.

Lokoja was a significant hub for the slave trade in the 18th and 19th centuries. The "Iron of Liberty," a representation of the efforts made by anti-slavery activists to free enslaved people, later recognized the location as a site of emancipation. This historical significance has had a lasting impact on the city, instilling pride among its residents. As a result, Lokoja is not just a site of remembrance but also a thriving center for cultural festivals that celebrate its diverse traditions and promote unity among its citizens.

Lokoja is an important hub for trade and a gathering place for agricultural goods such as kernels, cotton, and palm oil. The Kpata and New Markets are key commercial centers, offering a wide selection of regional foods and crafts. Minerals like iron ore and limestone are abundant in the towns surrounding Lokoja. This close proximity supports industrial operations such as the Obajana Cement Facility, Itakpe National Iron Ore Mining Company, and the nearby Ajaokuta Steel Company. The Ebira, Hausa, Igala, Nupe, and Oworo are among the many ethnic groups that enrich the city's cultural diversity, reflected in its local festivals, cuisine, and languages.

Religions such as Christianity, Islam, and traditional beliefs are practiced by the populace, and holidays such as Christmas and Eid are widely observed. A prominent natural landmark is the confluence point, where the Niger and Benue rivers meet. While the city features buildings from the colonial era, Mount Patti is a popular destination for breathtaking views of the city and its riverine landscape.

Historical sites include the earliest primary school in Northern Nigeria and the former residence of Lord Lugard, now the Kogi State Government House. Another historic monument marking the site of the emancipation of slaves is the Iron of Liberty. European troops and missionaries from the 18th and 19th centuries are interred in the European Cemetery, an historic cemetery in Lokoja.

4.12.5 Smart City Framework for Lokoja City

A Smart City framework for Lokoja should build on the city's potential as a growing urban hub while directly addressing its unique challenges, including its complex geography and rapid urbanization. To foster an inclusive, resilient, and sustainable urban environment, technology must be integrated with carefully targeted development priorities. The framework must tackle Lokoja's specific issues, such as fast population growth, inadequate infrastructure, and vulnerability to flooding. To meet these challenges effectively, the city should prioritize public-private partnerships (PPPs), context-appropriate technologies, and sustainable urban development in a practical, actionable plan.

4.12.6 Foundational Pillars of the Framework

A fundamental or foundational pillars diagram is an illustration of the guiding ideas or components that provide a solid foundation for a concept, objective, or organization. The roof or top structure, which represents the ultimate goal or vision, is typically supported by a number of vertical columns (the pillars) positioned on a common foundation. This type of diagram is used to depict core principles, strategic pillars, or the "building blocks" of success in fields such as business, management, or even democracy.

Robust governance and regulation

For the framework to be successful, it requires a comprehensive legal and regulatory structure that promotes transparency, protects data privacy, and facilitates coordination among various government entities. Effective project implementation also requires strong political will.

Community engagement

In low-income cities, there is often a lack of public awareness about Smart City initiatives. To ensure that solutions meet the needs of the community and build public trust, a successful framework must actively engage citizens through town hall meetings, public education, and community participation.

Integrated urban planning

According to previous research, Lokoja requires a new master plan to reorganize the distribution of facilities and manage its growing population. This strategy should integrate a Smart City framework as a core component of sustainable urban development rather than as an add-on technology.

Contextual technical solutions

Unlike frameworks designed for wealthy countries, Lokoja's approach should focus on implementing affordable, scalable, and flexible technologies that address pressing issues with substantial impact. Critical urban challenges can be tackled directly, for example, by using Internet of Things (IoT) solutions for waste management or real-time traffic monitoring.

Public-Private Partnerships (PPPs)

These partnerships are crucial for financing large-scale infrastructure and Smart City technology projects, as Nigerian cities often face severe financial constraints. Involving the private sector can help mobilize funds and provide technological expertise.

4.12.7 Key Dimensions of the Proposed Framework for the Lokoja's Smart City

Figure 4.13 illustrates the six fundamental elements that constitute the core dimensions of Lokoja's Smart City framework.

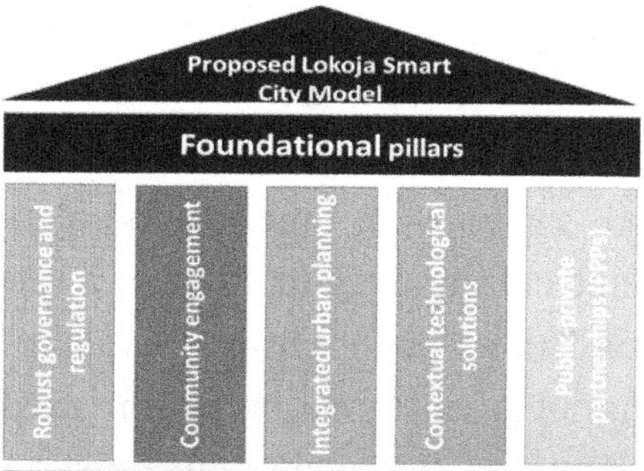

Figure 4.13: Fundamental Pillars of a Proposed Model Smart City for Lokoja, Kogi State, Nigeria

These elements are interrelated and collectively form the foundation for developing an inclusive, resilient, and sustainable urban environment, as further explained in the highlights below.

i. Smart environment

To protect Lokoja's ecology and improve residents' quality of life, an effective framework should promote sustainable practices. A smart environment—whether a smart home or a Smart City—uses interconnected sensors, devices, and software to create responsive, intelligent systems that adapt to the needs of people and the environment.

By seamlessly integrating technology to monitor, analyze, and manage resources and services such as energy, water, and transportation, a smart environment aims to enhance user experience, efficiency, and sustainability. This is enabled by the convergence of context awareness, intelligent systems, and ubiquitous computing, which together provide convenient, non-intrusive, data-driven services. The main characteristics are outlined below:

- **Flood resilience**: Lokoja is highly vulnerable to flooding due to its location at the confluence of two major rivers. Resilient infrastructure planning and early flood warning systems can be achieved through advanced GIS technology and community-driven architectural solutions. Real-time water level monitoring devices at the Niger-Benue confluence can provide timely flood alerts. These systems can leverage cutting-edge tools such as satellite imagery and Internet of Things (IoT) sensors to collect accurate data. By combining this information with predictive analytics, communities can effectively reduce risks and better prepare for potential flooding events.

- **Waste management**: Use IoT-enabled smart bins equipped with sensors to ensure timely waste collection, prevent overflow, and promote recycling. These smart bins can also help optimize collection routes and improve overall urban cleanliness.
- **Pollution control**: Install sensors to monitor air and water quality, providing data aligned with environmental standards. This proactive approach helps identify pollution sources and track improvements over time. By leveraging real-time data, policymakers can make informed decisions to protect natural resources and enhance public health.
- **Green spaces**: To address environmental challenges, prioritize the integration of eco-friendly urban designs and green spaces in city planning. Features such as green roofs, parks, and gardens not only enhance visual appeal but also improve air quality and support biodiversity. These initiatives can promote community well-being and foster stronger connections between residents and their surroundings.
- **Urban sprawl management**: Given that urban sprawl is a growing concern in Lokoja, remote sensing and GIS technologies can be used to monitor and manage it effectively. Strategies such as investing in public transportation, promoting higher-density housing, and enforcing zoning laws that encourage mixed-use developments can help curb sprawl. By implementing these measures, the city can reduce the environmental impact of urban expansion and foster sustainable growth.

ii. Smart Mobility

Smart mobility involves the use of data, technology, and sustainable alternatives to improve the environmental impact, safety, and efficiency of transportation. It integrates a range of physical solutions such as shared rides, electric vehicles, and public transport with digital solutions, including smart traffic management systems and Mobility-as-a-Service (MaaS) platforms, to make urban and non-urban transportation more efficient and accessible. Key objectives include reducing pollution, minimizing traffic congestion, and lowering dependence on fossil fuels. The following are the main components and concepts of smart mobility:

- **Intelligent transportation**: Deploy real-time traffic monitoring systems using IoT and data analytics to optimize traffic flow and reduce congestion, particularly on major routes such as Felele Main Road. These systems can improve commuter experience while lowering emissions by minimizing vehicle idle times. Additionally, integrating public transport with these technologies can further streamline urban mobility and promote sustainable travel options.
- **Public transit enhancement**: Introduce smart technologies to optimize public transportation routing and scheduling, providing efficient travel options and reducing reliance on private vehicles. By leveraging advanced infrastructure and digital solutions, public transit can become more convenient and attractive through increased service frequency, broader coverage, and greater reliability. Key areas of improvement include modern, environmentally friendly vehicles, AI-powered dynamic scheduling, enhanced accessibility for all users, integrated fare systems, and robust safety and security measures. These upgrades not only reduce dependence on personal vehicles but also generate economic and environmental benefits.
- **Smart parking**: Smart parking is an advanced system that simplifies and improves parking management by using sensors, cameras, and software. It allows facility owners to

manage operations, enforce regulations, and collect data, while providing drivers with real-time information about available spaces through mobile apps or digital signage, enabling faster and more convenient parking. This approach helps reduce traffic congestion, air and noise pollution, and the stress associated with searching for parking. There are two main technological options for implementing automated parking systems in commercial spaces: smart parking guidance systems and automated parking structures (APS). The choice between them depends on the specific needs, available space, and budget of each location.

iii. Smart and sustainable infrastructure

To manage Lokoja's rapid urbanization and unlock its full economic potential, it is essential to address the city's infrastructure gaps through smart and sustainable projects. Outdated infrastructure and planning have contributed to slum conditions, traffic congestion, and environmental challenges, but new government initiatives are beginning to pave the way for resilient, modern infrastructure. The key principles are outlined below:

- **Utility and power grid**: Develop smart grid technologies to integrate renewable energy sources and better manage Lokoja's inconsistent electricity supply. A multi-phase smart grid approach combining advanced management systems, improved grid efficiency, and renewable energy integration can help stabilize power delivery across the city. Beyond addressing local infrastructure and distribution challenges, this plan would leverage Kogi State's significant solar and hydropower potential.
- **Water management**: By integrating smart meters and sensors, Lokoja's water management system can monitor consumption, detect leaks, and improve supply reliability.

This IoT-based approach enhances efficiency and reduces water waste by measuring flow and pressure, immediately identifying anomalies, and sending real-time data to a centralized platform for analysis and rapid response.

- **Digital connectivity**: Developing Smart City services in Lokoja requires strong investment in reliable mobile internet and high-speed broadband. While Nigeria's digital infrastructure is being expanded through national initiatives, it is essential to localize these efforts in Lokoja to attract investment and ensure a robust digital foundation for future growth. The Federal Ministry of Communications, Innovation, and Digital Economy is actively leading this expansion. One notable initiative is Project Bridge, a public-private partnership (PPP) aimed at deploying a 90,000 km fiber-optic network nationwide. Additionally, private companies like IHS Towers are extending fiber-optic coverage and establishing rural connectivity sites across many Nigerian states.

- **Energy and utilities**: To enhance safety and promote environmental sustainability, smart energy solutions such as solar-powered street lighting should be implemented. These systems can also mitigate the challenges posed by unreliable power sources. Energy and utilities in Lokoja are overseen by the Kogi State Electricity Regulatory Commission (KERC) and the Abuja Electricity Distribution Company (AEDC), and the government is continuously working to improve supply through infrastructure upgrades and problem resolution. Water services are also improving, particularly following the 2022 floods, as the government rehabilitates and modernizes facilities. Despite these efforts, both water and electricity supply continue to pose challenges for residents, including billing issues, inconsistent availability, and the high cost of alternative power sources.

- **Transportation**: Implement an IoT-based traffic management system to address Lokoja's significant congestion problems. By leveraging real-time data, this system can optimize traffic flow and inform future transportation policies. Road transport comprising federal, state, and local highways is the primary mode of transportation in Lokoja, with tricycles and taxis commonly used for intra-city travel. Heavy reliance on federal routes contributes to severe traffic congestion. Rail and water transport are also important and should be revitalized to enhance connectivity and support economic growth. By strategically developing roads, waterways, and rail networks, Lokoja can become a regional transport hub, enabling efficient passenger and freight mobility.

iv. Smart Governance and Institutional Development

Smart governance is essential for a well-planned Smart City. In Lokoja, this involves modernizing and digitizing city management processes. By systematically transforming its administrative procedures, the city can foster smart governance and strengthen institutional capacity. The foundation for this digital shift was established in December 2024, when the Kogi State Government set up a Governance Delivery Unit (GDU) in Lokoja to focus on strategic planning, data analysis, and service delivery (Eneojoh, 2024). Effective smart governance in Lokoja requires the integration of technology, policy, and community participation to ensure transparency, efficiency, and inclusivity. The key elements are outlined below:

- **Provide a central authority**: Establish a Smart City board or specialized task force to oversee and coordinate all projects, ensuring alignment across government departments. The committee should include representatives from diverse sectors to enable a comprehensive approach to urban challenges. By fostering collaboration and communication among stakeholders, the task force can implement innovative solutions that enhance the quality of life for residents.

- **Adopt e-governance**: Replace manual, paper-based methods with digital platforms for public services to improve efficiency and expand citizen access to information. Incorporating data analytics can help identify community trends and challenges, enabling proactive solutions. Beyond streamlining processes, this modern approach empowers citizens to engage more actively with their local government.
- **Encourage public-private partnerships (PPP)**: Actively engage private investors and developers to finance infrastructure projects, potentially using models similar to Abuja's "Land Swap" program. This approach can improve the efficiency and effectiveness of project implementation by leveraging the resources and expertise of the private sector. By fostering a collaborative environment, cities can accelerate development and enhance public services for their residents.
- **Create a legal structure**: Develop a revised legislation that addresses both the opportunities and challenges of a Smart City, with a focus on innovation, governance, and community participation. Such a framework should encourage public-private collaboration and facilitate the adoption of advanced technologies. To ensure an inclusive approach to urban development, it is also essential to safeguard the needs and rights of all community members.

v. Smart economy and tourism

With its vast historical and geographical significance,

Lokoja has the potential to create a "smart tourism" economy by utilizing technology. This strategy leverages digital technologies to promote its rich history, enhance infrastructure, and improve visitor experiences all of which could draw in investment and generate employment. The following are the key components:

- **Geo-tourism**: Create a specialized mobile application for tourism that is powered by GIS and offers instant access to historical data and destinations. This would enhance the experience of tourists and promote the development of sustainable tourism. User-generated content can also be incorporated into the app, enabling travelers to submit their insights and suggestions. The platform can increase its value and highlight lesser-known locations by cultivating a community of active travelers, further encouraging safe travel practices.
- **Market development strategies**: By incorporating technology for automated billing, waste management, and security, the city's ultra-modern market, a component of its urban regeneration project, can be turned into a "smart" market and serve as a new center of the economy. Customers' shopping experiences will be improved by this innovation, helping draw in new companies and capital to the region. The market can serve as a template for upcoming urban projects in the area by creating a vibrant sense of community.
- **Inclusive growth strategy**: To guarantee that financial gains are distributed fairly, focus on using technology to assist small enterprises and urban agriculture, especially among the city's underserved populations. This strategy promotes community sustainability and resilience, in addition to empowering local business owners. By incorporating innovative solutions and providing access to

resources, we can build a more equitable economic environment that benefits everyone.

- **Opportunities for investment in smart infrastructure**: Smart infrastructure can be used to attract capital and support tech-based enterprises, thereby boosting employment and the economy. This strategy not only strengthens the local economy but also establishes the area as a center for creativity and innovation. Communities can build a resilient future that benefits both citizens and businesses by utilizing sustainable practices and state-of-the-art technology.
- **E-commerce strategies for medium and small-scale businesses for locals**: Encourage the growth of small enterprises by promoting digital platforms that extend trade and commerce beyond the city core. By connecting local customers with their favorite stores, this strategy not only increases visibility for these businesses but also fosters a sense of community. Their reach can be further expanded by utilizing social media and online marketing techniques, which can boost both foot traffic and online engagement.

vi. Smart People, Citizen Engagement, and Public Services

A people-centric approach that involves citizens as active participants in the development of the city must be given priority in the framework. The structure can ensure that public services are tailored to meet the specific needs of the populace by encouraging cooperation between public servants and community members. Ultimately, this interaction results in a more vibrant and resilient community, while also increasing transparency and fostering trust. The critical steps are outlined as follows:

- **Community participation**: Ensure that the community is actively involved in the development and implementation of smart city projects, as this can foster a sense of accountability and ownership. Additionally, use digital channels to engage citizens in the planning process and provide opportunities for ongoing public feedback on city services.

- **Digital literacy**: Provide initiatives to raise the general public's digital literacy, which is essential for a technologically dependent city. To ensure that residents can effectively use and benefit from emerging technologies, public education campaigns and skills-building programs should be launched.

- **Data integration**: Create a system for gathering and evaluating data in real time, which is currently absent, to continuously assess the city's performance and make informed decisions. A "Smart City Urban Data Integration Platform" can be used by cities to create a system for real-time data collection, integration, and analysis. Sensors, IoT devices, administrative databases, and social media feeds are just a few of the diverse urban data streams that these systems aggregate. Cities may use AI and machine learning to turn these fragmented systems into intelligent ecosystems that make data-driven decisions, continuously monitor performance, and proactively address evolving urban challenges.

The goal of creating a new Smart City framework for Lokoja is challenging but attainable. It calls for a multifaceted strategy that carefully balances technology with the city's unique socioeconomic and topographical circumstances. Lokoja may develop from a rapidly expanding city into a thriving, inclusive metropolis by emphasizing effective governance, robust infrastructure, an innovative economy, a sustainable environment, and active citizen participation.

4.12.8 Steps to implement a model of Lokoja's Smart City

To transform Lokoja into a Smart City, Kogi State must adopt a long-term, phased approach that emphasizes stakeholder collaboration, strong infrastructure, and technological integration. By utilizing technology, a smart city enhances its citizens' quality of life and optimizes urban services.

It is necessary to conduct a feasibility analysis and needs assessment in Lokoja by gathering information on the community's needs, existing infrastructure, and demographic trends through a comprehensive study. This will help tailor solutions to the city's specific needs. Initiate high-impact, small-scale pilot projects in selected areas to test technologies and demonstrate the benefits of the Smart City framework to stakeholders and the general public. Furthermore, establish a legal framework by enacting legislation and regulations that protect data privacy, promote smart city development, and clarify public-private partnership (PPP) guidelines. To fund the framework's implementation, a combination of government funds, international aid, and private sector investment through public-private partnerships (PPPs) will be used. Smart City initiatives should be implemented gradually, beginning with foundational projects such as smart grids and internet connectivity and progressing to more complex integrations like intelligent transportation systems. A continuous performance evaluation system should then be established, leveraging real-time data to monitor progress, make improvements, and ensure public accountability.

Phased Rollout

This provides a thorough foundation for creating a Lokoja Smart City project, as outlined below and illustrated in Figure 4.14.

i. Phase One - Planning and conceiving

Developing a shared, people-centered vision for Lokoja's

future that takes into account its unique requirements and challenges is the foundational stage.

- **Clearly state the vision**: Clearly define the goals for the Smart City transformation of Lokoja. The primary goals of the vision should be to improve the quality of life for residents, stimulate economic growth, and enhance public services.
- **Perform feasibility studies**: Evaluate the socioeconomic conditions, resources, and infrastructure that Lokoja currently has. Determine the population's particular demands as well as the difficulties facing the neighborhood.
- **Identify specific projects**: Evaluate the socioeconomic conditions, resources, and infrastructure that Lokoja currently possesses. Identify the population's specific needs, as well as the particular challenges facing the community.
- **Establish partnerships**: To attract capital and expertise, cultivate strong public-private partnerships (PPPs). These partnerships can ensure long-term sustainability and accelerate infrastructure projects.

ii. Phase Two - Building Foundational Infrastructure

The development of the fundamental digital and physical infrastructure that will underpin all Smart City projects is the main goal of this phase.

- **Boost digital connectivity**: Create a fiber-first digital infrastructure to provide reliable, high-speed internet access throughout the city. For smart devices and services to function effectively, widespread and affordable connectivity is essential.

- **Establish an open data platform**: To gather and examine data from sensors, devices, and user input, a secure, centralized data platform should be established. Ensure that data is handled responsibly and securely.
- **Create smart grids**: To monitor and optimize energy use, reduce waste, and promote renewable energy sources such as solar, the energy grid should be upgraded with smart meters and other advanced technologies.
- **Redesign waste management**: Install a smart waste management system that tracks fill levels and optimizes collection routes by using sensors in bins. In addition to saving costs, this will contribute to cleaner, more efficient cities.

iii. Phase Three - Execute targeted smart solutions

Lokoja can begin implementing technology-driven solutions to manage specific city operations now that the essential infrastructure is in place.

- **Develop smart transportation**: Develop intelligent traffic management systems to monitor and adjust traffic flow in real time, thereby reducing congestion and fostering smart transportation. For residents, real-time taxi and bus tracking can improve the efficiency of public transportation.
- **Boost public safety**: Install a network of CCTV cameras equipped with analytics tools for crime detection and real-time monitoring. The data can be leveraged by smart emergency services to identify the fastest routes to an incident.
- **Introduce e-governance services**: Create online platforms for administrative tasks, such as filing permits and paying taxes, and digitizing government services. Citizens will

benefit from enhanced convenience and increased transparency as a result.

- **Create smart environmental management**: Install environmental sensors to monitor air and water quality. Early warning systems can also be established for potential hazards, such as flooding, which is a concern in riverbank cities like Lokoja.

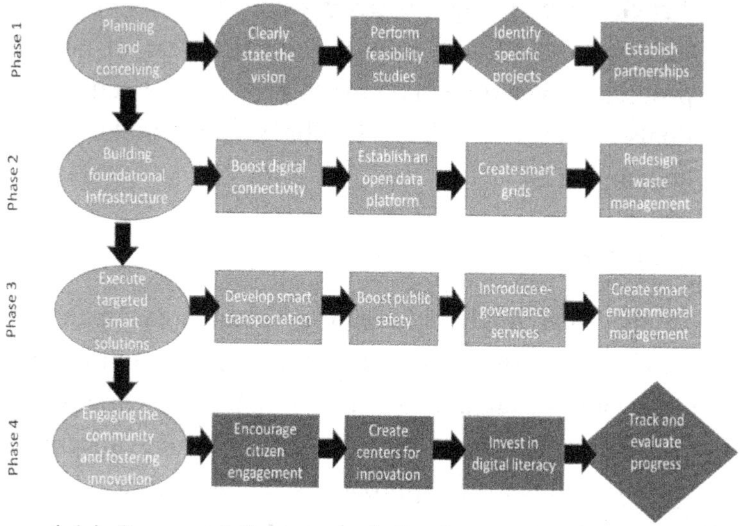

Figure 4.14: Proposed Framework for Smart City Implementation

iv. Phase Four - Engaging the community and fostering innovation

A Smart City's citizens are its most important asset. This final phase secures community buy-in and encourages a culture of continual innovation.

- **Encourage citizen participation**: Actively involve local companies, residents, and civic associations in the planning process. Organize workshops and town hall meetings to foster a sense of ownership.

- **Create centers for innovation**: Establish tech incubators to support new businesses and entrepreneurs in creating locally relevant solutions to urban challenges. To promote innovation, partner with academic institutions and research centers.

- **Invest in digital literacy**: Start initiatives to raise residents' digital literacy. This ensures that a large segment of the population can effectively use and benefit from emerging Smart City technologies.
- **Track and evaluate progress**: Use key performance indicators (KPIs) to monitor the progress of Smart City programs over time. To ensure projects are achieving their goals, gather resident feedback and make necessary adjustments.

4.13 Conclusions

In the final analysis, creating urban areas that are both technologically and financially efficient requires a framework of cost-effective and sustainable Smart City solutions. Cities can create a future that balances affordability and advancement by implementing innovative approaches in the areas of energy, transportation, waste management, and housing. In addition to improving residents' quality of life, this comprehensive strategy will build resilient communities capable of addressing challenges brought on by population growth and climate change.

Ultimately, cooperation between governments, corporations, and citizens working toward a common goal will be necessary for these projects to succeed. A shared objective that involves all parties in the decision-making process ensures that diverse viewpoints are considered, resulting in more inclusive and effective solutions. By cultivating a cooperative culture in all aspects, cities can use their combined knowledge and resources to create a sustainable and equitable future for everyone. In addition to fostering creativity, this cooperative strategy enhances resilience to the unpredictable impacts of climate change, enabling communities to flourish in the face of persistent challenges. Prioritizing education and awareness are crucial as we proceed, enabling people to participate in sustainability initiatives across all sectors. By cultivating a culture of participation, we can generate a sense of collective responsibility that encourages residents to actively engage in their local communities. Ultimately, this solidarity will result in stronger communities where everyone plays a vital role in creating a more equitable and sustainable society. The consequences for cities and their citizens are significant as urban planning continues to evolve due to technological breakthroughs. Large-scale data collection and analysis can enable better decision-making, but it also raises legitimate privacy and surveillance concerns among residents.

Furthermore, increased reliance on technology carries the risk of widening the digital divide and making it harder for underserved communities to access vital resources. To successfully address these challenges, urban planners must prioritize inclusive policies and promote meaningful community engagement. By ensuring that all voices are heard and considered during the planning process, cities can use technology not only as a tool for efficiency but also to advance equity and transparency for all residents.

Finally, this inclusive approach can lead to better-informed and more efficient decision-making, while also enhancing community ownership and trust in local government. Cities can ensure that diverse perspectives are considered when shaping their urban environment by leveraging data and communication channels. In addition to increasing civic engagement, this collaborative effort empowers residents to actively participate in the development of their city.

Local governments can then craft policies that reflect the interests and preferences of a broader population, ultimately resulting in more equitable and sustainable outcomes for all. These outcomes inspire residents to contribute to their communities, providing them with a sense of responsibility and belonging. Local governments can foster collaboration and trust by emphasizing responsiveness and inclusivity, thereby creating a foundation for a resilient and prosperous society.

"Smart Cities begin with smart choices" This statement implies that making intentional and well-informed decisions is essential to the creation of intelligent urban environments. Cities can improve their sustainability and functionality by emphasizing innovative approaches and efficient planning. These decisions span a range of areas, including waste management, energy consumption, and transportation. Residents benefit from a higher standard of living and more efficient use of resources as cities increasingly integrate technology into their infrastructure.

The development of smart cities depends on striking a delicate balance between innovative infrastructure integration and strategic decision-making. As smart solutions become more prevalent in urban settings, careful planning of waste management, energy use, and transportation networks becomes increasingly important for improving sustainability and functionality.

Cities can significantly enhance the standard of living for their citizens by utilizing technology to increase resource efficiency and streamline urban processes. Ultimately, the path to smarter urban living is paved with well-informed decisions that not only address current challenges but also anticipate future needs, resulting in thriving communities that are resilient and innovative.

References

Ali, A. D., & Nomishan, T. S. (2024). Accessibility: The Role of Transportation on Urbanization in Lokoja, Nigeria, 1854-1991. *Romanian Journal of Historical Studies*, *7*(1-2), 45-54.

Allam, Z., & Dhunny, Z. A. (2019). On big data, artificial intelligence and smart cities. *Cities*, *89*, 80-91.

Angelidou, M. (2017). The role of smart city characteristics in the plans of fifteen cities. *Journal of Urban Technology*, *24*(4), 3-28.

Aondofa, C. A. (2021). Green Area Mapping of Ahmadu Bello University, main campus, Samaru – Zaria, Nigeria https://www.theabusites.com/founder/chila-andrew-a/

Apata, Z. O. (1990). Lugard and the Creation of Provincial Administration in Northern Nigeria 1900-1918. *African study monographs*, *11*(3), 143-152.

Bello, E. I., & Usifo, C. I. (2009). A study of urban traffic management-A case study of Lagos State traffic management authority. *Advanced Materials Research*, *62*, 599-605.

Caragliu, A., & Del Bo, C. F. (2019). Smart innovative cities: The impact of Smart City policies on urban innovation. *Technological Forecasting and Social Change*, *142*, 373-383.

Chen, L., & Ma, R. (2024). Clean energy synergy with electric vehicles: Insights into carbon footprint. *Energy Strategy Reviews*, *53*, 101394.

Du, Y., Sun, T., & Huang, Y. (2015). A Comprehensive Review of Smart City Initiatives in Singapore Technological Advancements and Urban Planning from 2015 to 2023.

Elkington, J. (1994). Towards the sustainable corporation: Win-win-win business strategies for sustainable development. *California management review*, *36*(2), 90-100.

Eneojoh, A. B. (2024). Governor Ododo Unveils Governance Delivery Unit to Transform Kogi State.

https://kogistate.gov.ng/governor-ododo-unveils-governance-delivery-unit-to-transform-kogi-state/#:~:text=Transform%20Kogi%20State*Governor%20Ododo%20Unveils%20Governance%20Delivery%20Unit%20to%20Transform%20Kogi%20State,excellence%20and%20resilience%20in%20governance.

Ferro-Escobar, R., Vacca-González, H., & Gómez-Castillo, H. (2022). Smart and sustainable cities in collaboration with IoT: The Singapore success case. In *Machine Learning for Smart Environments/Cities: An IoT Approach* (pp. 213-243). Cham: Springer International Publishing.

Gowda, V. D., Prasad, K. D. V., De, T., Srinivas, V., Kumar, N. A., & Moharekar, T. T. (2023, April). A framework for smart city implementation using IoT-cloud based system architecture. In *2023 International Conference on Distributed Computing and Electrical Circuits and Electronics (ICDCECE)* (pp. 1-6). IEEE

Holland, O. (2021). Singapore is building 42,000-home Eco 'Smart' City https://edition.cnn.com/style/article/singapore-tengah-eco-town

Hollands, R. G. (2015). Critical interventions into the corporate Smart City. *Cambridge journal of regions, economy and society*, *8*(1), 61-77.

Kaenzig, R., Mobereola, D., & Brader, C. (2010). Africa's first bus rapid transit system. *Transportation research record, 2193*(1), 1-8.

Khan, J., & Labonté, O. (2021). The need for Smart City FRAMEWORKS. *Economic Development Journal*.

Khern, N. (2019). Digital government, smart nation: Pursuing Singapore's tech imperative. *Civil Service College Singapore*.

Kwangtae, Y. (2019). What is Smart Water Management? https://development.asia/explainer/what-smart-water-management

Prateeppornnarong, D. (2025). Cities and sustainability: Exploring contributions, opportunities and challenges of Smart city implementation towards social sustainability. *Urban Governance*.

Salawu, Y. 0. (2024). Development of Coastal Economies in Nigeria: A sustainability concept. 8/9 November, 2024 by Nam Can Tho University (NCTU) & Malaysia University of Science and Technology (MUST), published by Malaysia University of Science and Technology (MUST). NEW-ICDI-7.1-2024_Conference-eProceedings_251124-1.pdf. Pg. 77-87.

Sennett, R. (2012). No one likes a city that's too smart. *The guardian, 4*. December, 2012. http://www.guardian.co.uk/commentisfree/2012/dec/04/smart-city-rio-songdo-masdar

Umoh, A. A., Ohenhen, P. E., Chidolue, O., Ngozichukwu, B., Fafure, A. V., & Ibekwe, K. I. (2024). Incorporating energy efficiency in urban planning: A review of policies and best practices. *Eng. Sci. Technol. J, 5*, 83-98.

Wilson, R. (2020). The Journal of Lieutenant Charles Knowles in the River Niger, 1864. In *The Naval Miscellany* (pp. 273-325). Routledge.

Yang, R., & Zhen, F. (2024). Smart City development Models: A cross-cultural regional analysis from theory to practice. *Research in Globalization, 8*, 100221.

CHAPTER 5
Institutional Capacity: Resilience Building

5.1 Introduction

Resilience is the ability of social, economic, environmental, and institutional systems to absorb, adapt to, and recover from disturbances while maintaining core functions. In the 21st century, the need for sustainable development, resilience, and adaptability has never been more urgent. Across the globe, nations are facing interconnected challenges, including climate change, pandemics, geopolitical conflicts, urbanization, and rapid technological change. In Nigeria, a nation of over 200 million people and projected to be the third most populous country by 2050, these challenges are particularly pressing.

The global trajectory depends on the ability to plan tomorrow while confronting the realities of today. Resilience is not merely a buzzword; it is a strategy for national survival. Planning for resilience involves designing systems that can adapt, absorb, and recover from shocks while promoting social inclusion, economic growth, and environmental sustainability.

Long-term planning is a form of strategic foresight that involves policy development spanning decades, including anticipating future challenges, defining sustainable development goals, and designing frameworks for effective resource allocation. Global trends such as rapid population growth, resource scarcity, and ongoing socio-economic instability make this forward-thinking approach even more urgent.

This section examines the critical importance of long-term planning, urban resilience, and the development of Smart Cities as key strategies for securing the future of global societies in an increasingly dynamic world. There is growing concern regarding the capacity of Smart City development to safeguard environmental sustainability. In addition, numerous studies have reviewed the state of the art in Smart City development and governance. Some of these reviews focus on the application of big data and artificial intelligence in Smart Cities and their potential to promote environmental sustainability, as well as the technological risks associated with IoT-based systems. Other studies place greater emphasis on governance frameworks, policy coordination, and institutional challenges in Smart City implementation.

The latter reviews have attempted to identify the drivers behind smart cities, develop performance assessment frameworks for smart cities, clarify the desired outcomes of smart cities, and examine the contextual factors that influence both the components and the desired outcomes.

These comprehensive reviews, which explain how smart cities are planned, envisioned, and implemented in developed contexts, have laid the foundation and defined the objective of this review, which explicitly focuses on consolidating evidence related to smart city development in developing countries.

We are specifically interested in understanding the current state of Smart City development in developing countries, the unique set of driving factors that propel the emergence of smart cities in these contexts, and the distinct challenges that developing countries face when embarking on diverse Smart City initiatives. Despite extensive adoption efforts, the Smart City concept remains largely opaque even in developed countries.

The governance of Smart Cities presents additional complexities for many developing countries, particularly those still striving to meet the basic needs of their populations. Unlike developed nations, where foundational infrastructure and services are largely in place, developing countries face distinct governance challenges that further complicate the implementation of Smart City initiatives. These differences highlight the critical need for a systematic review of existing literature to better understand the context-specific dynamics, opportunities, and constraints surrounding Smart City development within developing countries. Furthermore, increasing rural-to-urban migration is expected to generate greater ecological, social, and infrastructural pressures in many major urban centers of developing countries.

As of 2018, approximately 55% of the global population resided in urban areas, and this trend is expected to continue rising over the next few decades. By 2030, most of the 43 megacities (cities with populations of 10 million or more) are expected to be located in developing countries. Nevertheless, the high financial costs associated with infrastructure maintenance, along with the substantial size of informal economies in developing countries, pose significant challenges to government-led Smart City aspirations, which require closer examination.

Moreover, privileging technology as the core of Smart City development without aligning it with public values or understanding the scope of citizens' basic needs could transform smart cities into mere "white elephants." Given the need to balance social needs with technological development in developing countries, enhancing the understanding of how smart cities can be governed amid competing developmental challenges become increasingly imperative.

In the contemporary era, cities' adoption of information-driven and environmentally conscious principles has transformed them into hubs of population growth and knowledge expansion at an accelerating pace. The potential of advanced Information and Communication Technology (ICT) to foster a sustainable future is most clearly demonstrated in these cities.

The convergence of efforts to integrate eco-friendly urban neighborhoods with intelligent urban systems has been driven by numerous technological and policy advancements. Amid the era of big data, there is a prevailing belief in the transformative power of innovative urban strategies, technological progress, and institutional reforms in shaping intelligent and sustainable urban processes. Therefore, policymakers should prioritize investments in data infrastructure to enable the collection, analysis, and application of big data for sustainable urban development. This will require efforts to strengthen data collection processes, facilitate data exchange among stakeholders, and allocate resources to technologies for data processing and analysis.

Policies should encourage the integration of data-driven approaches into urban planning. This involves incorporating data analytics tools and methods into decision-making processes, ensuring that urban development plans are guided by evidence-based insights on sustainability.

Policymakers should allocate resources to capacity-building and training programs to equip relevant stakeholders with the skills and knowledge necessary to effectively leverage big data for sustainable urban development. One approach is to provide training in data analytics, data visualization, and other pertinent areas to urban planners, policymakers, and related professionals.

To date, numerous case studies aimed at developing theoretical frameworks for Smart City governance have been conducted and published, particularly in developed economies. These empirical studies range from those examining implementation experiences, policy mechanisms, and transformation processes of Smart Cities to those that highlight the drivers and barriers, as well as studies proposing new metrics for performance evaluation, sustainability assessment, and the benchmarking of Smart Cities.

Nigeria, the most populous country in Africa, faces unique developmental challenges, including poor infrastructure, limited access to healthcare and education, and economic constraints. However, Nigeria also possesses enormous potential, driven by a young population, a developing digital economy, and an innovative spirit. Emerging technologies, such as the Internet of Things (IoT) and artificial intelligence (AI), are transforming industries globally and offering unprecedented opportunities to advance beyond traditional development pathways. While the journey toward technology-driven transformation in resource-constrained environments like Nigeria is undoubtedly challenging, it remains full of promise. By actively engaging key stakeholders, including local communities, entrepreneurs, and policymakers, it is possible to co-create scalable and inclusive solutions that align with Nigeria's unique socioeconomic realities.

5.2 The Engine of Smart City Implementation

Institutional capacity refers to the ability of public institutions to plan, implement, govern, and sustain development projects effectively and efficiently. It encompasses a range of skills, including technical competence, organizational structure, financial management, human resources, and operational effectiveness, that enable institutions to fulfill their mandates successfully.

For Smart Cities, institutional capacity serves as the foundation upon which policies and plans are translated into tangible outcomes. For smart city initiatives to succeed, institutions at all levels must possess the capability to manage complex, technology-driven urban systems. This includes the ability to conduct needs assessments, procure and implement technology solutions, facilitate inter-agency coordination, regulate private sector actors, and evaluate project performance.

Institutions in many low-income cities are currently hampered by limited technical capacity, constrained financial resources, bureaucratic inefficiencies, and widespread corruption (Angelidou, 2017). For instance, energy, environmental, and urban transport agencies often operate in silos, with overlapping mandates, weak coordination, and outdated processes. Accordingly, intelligent smart solutions, such as traffic lights or GIS-based waste management monitoring systems, are often inadequately integrated or entirely absent (Abubakar & Aina, 2019; Al-Kodmany, 2018; Alsaid, 2021). In other cases, digital products are procured without sufficient capacity-building for staff or sustainable maintenance plans, leading to rapid obsolescence (Meijer & Bolívar, 2016). To strengthen institutional capacity, the nation and other developing regions should invest in regular staff training, digitize administrative processes, and enhance inter-agency coordination. This also includes improving public financial management to ensure transparency and accountability in budgeting and the procurement of Smart City technologies.

In addition, partnerships with research institutions and technology parks can foster a knowledge ecosystem that supports local innovation and problem-solving. Ultimately, without strong institutions to guide and sustain digital innovations, even the most advanced Smart City technologies will not yield lasting impact. Institutional capacity building is therefore not merely a technical requirement but a strategic imperative for realizing Smart City ambitions in Nigeria. Institutional capacity refers to the ability of organizations, particularly public agencies, to execute tasks, solve problems, and set and achieve goals (Ateş, 2024; Azevedo et al., 2018; Albino & Dangelico, 2015; Belsky, 2012).

5.3 Governance Policy

These three foundations governance, policy, and institutional capacity are highly interdependent. Weaknesses in any one of these areas can undermine overall Smart City performance. For example, innovative policies may be ineffective if institutional capacities are inadequate, or government failures can lead to corruption and inefficiency despite well-designed policies.

A systems-based approach that integrates all these components promotes resilience, diversity, and creativity. The Smart City model of Singapore demonstrates how centralized governance, effective policy leadership, and robust institutions can coalesce to produce world-class outcomes (Chong, 2021). Smart Cities represent a new revolution in urban thinking, offering a way to apply digital technology to deliver enhanced services, improve the quality of life, and foster sustainable economic growth. As emphasized in this section, the success of Smart Cities depends not only on technological adoption but also on strong governance, coherent and adaptable policy frameworks, and the institutional capacity to implement and sustain innovation.

Globally, the need for Smart City adoption is underscored by rapid urbanization, growing infrastructural deficits, environmental degradation, and rising socio-economic inequality in cities (Cartwright et al., 2018). Good governance is essential to coordinate stakeholders, enhance citizen participation, ensure accountability, and prevent the exclusionary effects of top-down planning. Similarly, sound policies should serve as the legal and strategic frameworks that enable innovation while safeguarding the public interest.

Meijer & Bolívar (2016). Equally important is the capacity of institutional bodies to manage Smart City projects from conceptualization through deployment and maintenance, supported by qualified personnel, adequate funding, and transparent procedures. Finally, to move from rhetoric to action as drivers of inclusive urban transformation, national and subnational governments must recognize digital innovation as a governance challenge rather than solely a technological issue.

This entails political will, sustained investment, and collaborative partnerships with the private sector, civil society, and academic institutions. When these conditions are met, low-income countries can unlock the full potential of smart urbanism, creating cities that are not only smart but also resilient, inclusive, and future-ready (Colding et al., 2024). Forward-thinking governance is essential for the adoption of emerging technologies. Nigeria must establish laws that ensure user data protection, transparency in AI decision-making, and accountability on digital platforms. Although the Nigeria Data Protection Regulation (NDPR) is a strong starting point, more robust frameworks are needed, particularly regarding cybersecurity and AI ethics. End users, industry, academia, and civil society should all be actively involved in policy co-creation.

Using AI ethically also entails avoiding algorithms that reinforce social inequities and ensuring diversity and representation in datasets. A better-informed and secure populace can be achieved through public education on digital rights and responsibilities (Castelnovo, Misuraca, & Savoldelli, 2020).

5.4 Governance, Policy, and Institutional Capacity: Pillars of Sustainable Smart City Initiatives

In response to rapidly growing populations, cities are expanding, which intensifies pressure on governments worldwide to provide effective solutions. To improve infrastructure, deliver services seamlessly, and enhance the quality of urban life, many governments are relying on smart technologies. However, the concept of a Smart City encompasses far more than the mere use of technology; it requires a coherent development and growth strategy that prioritizes inclusivity and safeguards the well-being of citizens.

As the need for such transformation becomes critical, urbanization in the metropolitan region has significantly slowed economic growth. In addition to these factors, a lack of effective governance results in chaotic traffic, power outages, growing waste, dwindling water supplies, and declining public safety. These problems are exacerbated by weak governance, poor policy coordination, and an inability to manage overextended bureaucracies (Healey et al., 2017; Renda, 2019).

Adopting a Smart City framework globally provides decision-makers with the opportunity to overcome some of the traditional constraints that have hindered progress. With innovations such as digital traffic management, electronic utility billing, and e-government services, municipal governments can enhance service delivery, streamline urban planning, and strengthen public trust. However, these innovations must be tailored in the context of developing countries to succeed. Implementing models from more developed nations can be challenging for some African countries, where a significant portion of urban residents live in informal settlements, digital skills are limited, and internet access is not universal.

In this context, low-income countries need to develop a Smart City strategy that works within local realities, leverages community intelligence, mitigates social inequities, and bridges the digital divide. Such an approach also involves formulating policies that are transparent, clear, enforceable, and respectful of citizens' rights in an era of heightened digital surveillance. Developing Smart Cities goes beyond technology; it also requires strong leadership, well-defined policies, and agile institutions capable of delivering urban transformation.

According to scholars studying low-income countries, innovation and development are ongoing processes that support the integration of Smart Cities, while governance, policy, and institutional capacity shape cognitively innovative Smart City projects (Ismagiloiva et al., 2019; Kesar & Ache, 2024; Kirimtat et al., 2020). The essential infrastructure of Smart Cities, including digital technology, the Internet of Things (IoT), artificial intelligence (AI), and big data, is defined by the use of modern technologies. Cities can monitor resources, optimize services, and proactively address the needs of their inhabitants thanks to these tools. The transition from traditional reactive systems to proactive, efficient, and resilient urban governance is demonstrated in the integration of smart grids, intelligent transportation, e-governance, and data-driven healthcare systems.

5.5 Emerging Technologies

According to UN-Habitat (2020) and the World Economic Forum (2020), "emerging technologies are technologies whose development, practical applications, or both are still largely unrealized." Global change is being accelerated by emerging technologies such as blockchain, the Internet of Things (IoT), and artificial intelligence (AI). AI refers to systems that can perform tasks such as pattern recognition and decision-making that normally require human intelligence.

Smarter environments are enabled by the Internet of Things (IoT), which consists of interconnected devices that collect and share data. Blockchain, in turn, provides transparent and secure methods for recording transactions. Despite their global momentum, adoption in low-income countries is hampered by high costs, inadequate infrastructure, and limited digital literacy (Daramola, 2024; Dierwechter, 2017).

Figure 5.1: The Use of Modern Technology

To fully leverage these technologies in such contexts, a comprehensive understanding is essential. In many parts of the world with constrained resources, particularly low-income countries with underdeveloped infrastructure, the challenges are real and pressing, even though wealthier nations have already integrated these technologies into everyday practice. These challenges include insufficient policy frameworks, low digital literacy, limited financial capacity, and inadequate infrastructure. However, these difficulties also present an opportunity: developing countries can adopt affordable, flexible, and agile technologies tailored to their unique contexts by moving beyond outdated systems (Alahi et al., 2023).

With the increasing urbanization of the world, inclusive urban transformation has become a critical approach to urban development today. Identifying effective models for sustainable urban development is the first and most crucial step in enabling Smart City development.

The so-called inclusivity development model is an economic growth framework focused on long-term development, in which indicators of urban development are grounded in sustainable development theory. This model has significant theoretical and practical value, as it can accelerate sustainable development, ensure sustained and rapid economic growth, promote comprehensive, coordinated, and sustainable progress, and achieve inclusive urban growth.

Earlier studies examined the sustainable development of cities from the perspectives of transport systems, energy sustainability, and land use (Gunnarsdóttir et al., 2020; Llorca et al., 2020; Karjalainen & Juhola, 2021). Although the indicator system for sustainable cities is being developed gradually, how it can scientifically evaluate different types of cities and how to enhance the flexibility and stability of the system remain areas for further research.

Based on the evaluation of a wide range of areas across the country, a scientific and rational framework aligned with the concept of sustainable urban development is established using five dimensions: economic development, social and living conditions, resources, environment, and policy.

Factors such as the intensity of urban economic development, industrial structure, and international urban integration are considered to promote more equitable urban development (Domorenok et al., 2021; Greera, 2022; Lee & Park, 2025). Figure 5.2 illustrates that the health benefits of living and working in Smart Cities reinforce the connection between smart infrastructure and public health.

Figure 5.2: Health Benefits of Working at Smart Cities
(World Health Organization, 2025 Health Manual Book)

By reducing pollution through smart transportation, optimizing energy consumption, and improving waste management, Smart Cities create cleaner and healthier environments. Moreover, digital health monitoring and telemedicine are making healthcare more accessible, particularly in urban areas where health disparities persist. A healthy city is a sustainable city; therefore, smartness serves as a pathway to human well-being and survival.

5.7 Inclusivity and Marginalized Groups

Urbanization in developing countries is occurring at an unprecedented rate, often resulting in spatial inequality, exclusion, and the marginalization of vulnerable populations. Inclusive planning has emerged as a critical framework to ensure that urban growth benefits all residents equitably, particularly the poor, women, persons with disabilities, and other marginalized groups (Singla et al., 2024; UN-Habitat, 2023). Inclusive urban planning ensures that development processes reflect the diverse needs of urban populations, addressing not only infrastructure gaps but also systemic social exclusion. In developing cities, where informal settlements and economic inequality are widespread, inclusive planning fosters social cohesion, economic opportunity, and sustainable urban growth (UN-Habitat, 2020). Without such approaches, rapid urbanization often leads to deepening inequality, unplanned urban sprawl, and increased vulnerability to climate-related risks.

Effective inclusive planning begins with enabling meaningful citizen participation, particularly among marginalized groups. Participatory mechanisms such as community mapping, public consultations, and citizen assemblies can empower residents to influence planning decisions (Lyu, Lu, Lee, Schmitt, & Lim, 2022). Moreover, digital technologies—such as mobile applications for feedback and SMS-based surveys—offer scalable and cost-effective ways to involve low-income populations in urban planning (World Bank, 2015; Lyu et al., 2022).

Legal frameworks often fail to recognize informal settlements, thereby preventing residents from accessing services or participating in planning processes. Reforming land-use policies, formalizing tenure rights, and adopting inclusive zoning laws are therefore critical steps (Verhulst & Young, 2017). Decentralization and the strengthening of municipal institutions also enable more localized and inclusive decision-making. Inclusive planning in developing cities is both a social imperative and a practical necessity.

It requires political will, cross-sector collaboration, and continuous engagement with urban communities. While challenges remain such as institutional inertia and resource constraints examples from various cities show that inclusive planning is achievable and transformative when approached holistically. Advancing inclusive urban planning in developing cities requires a multifaceted approach that integrates participatory governance, robust legal frameworks, data-driven planning, inclusive infrastructure development, economic inclusion, and climate resilience. Recent studies and initiatives underscore the importance of adopting inclusive strategies to address urban challenges and promote sustainable development. By prioritizing inclusivity in urban planning processes, cities can foster more equitable and resilient urban environments for all residents (UNESCO, 2023).

5.8 Perspective on Developing and Low-Income Countries and Sustainability Scale on Smart City Efforts

The notion that a city could be smart was largely an imagined concept in popular culture for much of the 20th century, but with the widespread adoption of devices capable of computing at multiple scales and possessing embedded intelligence, the possibility that a city could become intelligent even self-aware is now considered a new standard.

Aimed at providing real-time data on the movement of people and materials, as well as supporting decision-making processes related to the physical and social structure of the city, Smart Cities are typically envisioned as constellations of devices at multiple scales connected by various networks.

Furthermore, Smart City projects continue to contribute to infrastructure improvement, system digitization, service integration, platform automation, safety, stakeholder engagement, and meeting citizen expectations because they leverage knowledge and interactive technologies to enhance economic development, environmental sustainability, individual well-being, and service quality.

According to global market research by Navigant Research (2017), 178 cities worldwide had deployed over 250 Smart City projects as of 2017. Similarly, with 12 cities ranked among the top 25 Smart Cities in the IESE Cities in Motion Index for 2018, Europe emerges as a global leader in Smart City development (IESE, 2018). The complexity of Smart City components and the challenges they present has prompted plans to adopt multidisciplinary approaches and integrate insights from multiple fields of inquiry (Patrão et al., 2020; Shamsuzzoha et al., 2021).

Considering this information, the most prevalent and significant objectives of Smart City initiatives are sustainability across society, the environment, and the economy. Future urban development planning must take social sustainability into account (Monfared Zadeh et al., 2015).

In Smart Cities, cultural diversity can pose challenges for both individuals and communities. Hence, there is growing interest in expanding Smart City evaluations to incorporate various socio-technical perspectives and interconnections (Visvizi et al., 2018). Cappai et al. (2019) developed efficient processes for sustainability assessment and produced indicators based on socioeconomic and environmental parameters.

Lowery et al. (2020) propose an assessment index system for sustainable cities and consider regional sustainable development as well as the influence of natural resources, particularly in the forestry and fisheries sectors. The Smart City has become a popular political concept, widely embraced as an ambitious vision of development in many countries. However, Smart City development is neither new nor recent (Kummitha & Crutzen, 2017).

Figure 5.3 shows a compiled illustration of Smart City components, offering a comprehensive view of key domains that define a Smart City, including mobility, governance, environment, economy, and living standards.

This integrated model underscores the interdependence of various urban functions and demonstrates how their smart optimization contributes to urban resilience. For instance, intelligent traffic systems not only reduce congestion but also lower carbon emissions, enhancing the city's ecological sustainability.

Figure 5.3: Compiled List of Smart City Illustration

5.9 Smart City in Nigeria – A Case Study

Smart Cities leverage technology to enhance urban living, but ensuring inclusivity remains a challenge. Nigeria's urban population is growing rapidly, with estimates suggesting that the country will be home to several megacities by 2030. The concept of Smart Cities has gained global popularity, with many cities embracing technology-driven solutions to improve infrastructure, public services, and overall quality of life.

However, the benefits of Smart Cities are not always equitably shared, and marginalized groups often face significant barriers to accessing them. In Nigeria, the lack of inclusivity in Smart City development can reinforce existing social and economic inequalities, limiting the potential of Smart Cities to promote sustainable and equitable development.

Nigeria's urbanization challenges, including inadequate infrastructure, traffic congestion, and insecurity, can be addressed through Smart City initiatives, provided these initiatives are designed and implemented to prioritize inclusivity and accessibility. While rapid urbanization presents opportunities for smart city development, it also risks exacerbating existing social inequalities.

The country's urban population is projected to continue growing, placing pressure on cities to provide essential services such as healthcare, education, and sanitation.

Smart City initiatives offer tools to address these challenges and can promote economic growth, improve quality of life, and enhance citizen engagement, but only if they are designed to benefit all segments of society. Urban development plans and policies must therefore prioritize inclusivity and sustainability to ensure that Smart City initiatives meet the needs of all citizens.

Figure 5.4: Nigeria cityscape (Lagos)

Marginalized groups such as women, persons with disabilities, and low-income communities are often excluded from Smart City benefits, thereby perpetuating social and economic inequalities. In Nigeria, these groups face significant barriers to accessing technology, education, and economic opportunities, which can hinder their meaningful participation in Smart City development.

The exclusion of marginalized groups from Smart City benefits can have far-reaching consequences, including reduced access to healthcare, education, and employment opportunities. This can entrench cycles of poverty and inequality, limiting the potential of Smart Cities to promote sustainable and equitable development. Furthermore, a lack of inclusivity in Smart City development can lead to social unrest and conflict, as marginalized groups may feel that their needs and concerns are not being addressed. The main purpose of this case study is to investigate how Smart City initiatives can be designed to include marginalized groups and to identify strategies for bridging digital and social gaps.

It also aims to explore the potential of Smart Cities to promote inclusive development and determine ways to ensure that marginalized populations are not left behind. Additionally, the study examines the experiences of marginalized groups in Nigeria's Smart Cities and identifies best practices for inclusive Smart City development. It further explores the role of policy and regulatory frameworks in promoting inclusivity and identifies opportunities for capacity building and stakeholder training.

5.9.1 Framework for Inclusive Smart City Development

Participatory urban planning, human-centered design, and inclusive policymaking can help ensure that Smart City initiatives benefit all segments of society. These frameworks prioritize the needs and experiences of citizens, making Smart City initiatives more responsive to marginalized groups. Other frameworks, such as the Capability Approach and the Human Development Index, can provide valuable insights into the design of inclusive Smart Cities. Moreover, these frameworks can assist policymakers and stakeholders in evaluating the effectiveness of Smart City initiatives in promoting inclusivity and identifying areas for improvement.

5.9.2 Nigeria's Smart City Landscape

Nigeria has initiated several Smart City projects, but ensuring inclusivity remains a significant challenge. The country's Smart City initiatives have focused on improving infrastructure, services, and quality of life, yet they often lack inclusive design and implementation. This can restrict the benefits of Smart City development for marginalized groups, perpetuating existing social and economic disparities. Nigeria's Smart City initiatives must prioritize inclusivity and accessibility to ensure that all citizens can benefit from urban development. The government and stakeholders must collaborate to develop policies and regulations that promote inclusivity, accessibility, and affordability in Smart City planning and implementation.

The country has launched several Smart City initiatives, including the Lagos Smart City Initiative, the Abuja Smart City Project, and the Nigerian Smart City Strategy. These initiatives aim to improve infrastructure, services, and quality of life, but they must be designed and implemented to prioritize inclusivity and accessibility. Nigeria's Smart City initiatives, however, have faced several challenges, including limited funding, inadequate infrastructure, and insufficient inclusivity.

The country's Smart City programs must also address the needs of marginalized groups, including women, persons with disabilities, and low-income communities. Nigeria's Smart City initiatives offer significant opportunities for promoting inclusive and sustainable development. These initiatives can help improve the quality of life for all citizens, foster economic growth, and enhance citizen engagement.

5.9.3 Locations of Case Study Cities

Lagos, Abuja, and Port Harcourt provide valuable insights into Smart City development and inclusivity challenges in Nigeria. These cities have implemented various initiatives, including intelligent transportation systems, smart lighting, and digital governance platforms. However, these initiatives often lack inclusive design and implementation, limiting their benefits for marginalized groups. The case studies will highlight both the challenges and opportunities for inclusive Smart City development in Nigeria.

Lagos, one of Nigeria's most populous cities, has implemented several Smart City initiatives, including the Lagos Smart City Initiative. The initiative aims to improve infrastructure, services, and quality of life, but it must prioritize inclusivity and accessibility to ensure that all citizens benefit. Similarly, Abuja, Nigeria's capital city, has launched several Smart City initiatives, including the Abuja Smart City Project.

This project aims to improve infrastructure, services, and quality of life, but it must address the needs of marginalized groups to ensure inclusiveness. Moreover, Port Harcourt, a major city in Nigeria's oil-producing region, has implemented several Smart City initiatives, including smart lighting and digital governance platforms. However, these initiatives must prioritize inclusivity and accessibility to ensure that all citizens benefit. By examining these case studies, we can identify both best practices and challenges in inclusive Smart City development in Nigeria and develop recommendations to promote inclusivity in Smart City initiatives.

Figure 5.5: Nigeria Smart Cities Initiative

Figures 5.4 and 5.5, depicting Nigeria's cityscape, Lagos, and the Nigeria Smart Cities Initiative, contextualize the global Smart City narrative within Nigeria's urban development trajectory.

Lagos, Nigeria's most populous city, represents both the opportunity and urgency for Smart City development. As urban population growth outpaces infrastructure expansion, the Nigeria Smart Cities Initiative seeks to introduce data-driven planning, intelligent services, and inclusive urban innovation. This strategy is not only about modernity but also about survival ensuring that cities like Lagos remain livable amid rising urban pressures.

In conclusion, Smart Cities are not simply a luxury or a vision of technological futurism; they are essential for survival and sustainable development. Figures 5.1 to 5.5 collectively demonstrate that building cities that endure requires aligning smart innovations with health, environmental, economic, and social resilience. Investing in smart city frameworks is an investment in the long-term sustainability and continuity of urban life.

Existing initiatives often prioritize technological innovation over social inclusion, underscoring the need for more inclusive approaches. Analysis of Nigeria's Smart City landscape reveals significant gaps in inclusivity, with marginalized groups frequently excluded from the benefits of these initiatives. Such exclusion can entrench existing social and economic inequalities, limiting the potential of Smart Cities to promote sustainable and equitable development.

5.9.4 Marginalized Groups in Nigeria's Smart Cities Women and Girls

They face challenges in accessing technology, education, and economic opportunities, which can limit their participation in Smart City development. Specific issues include:

- Limited access to digital literacy programs and training
- Gender-based violence and harassment in public spaces
- Unequal representation in decision-making processes

People with Disabilities

They require accessible infrastructure, services, and opportunities to participate effectively in Smart City development. Specific issues include:

- Inaccessible public spaces and transportation systems
- Limited access to information and communication technologies
- Lack of representation in decision-making processes

Low-Income Communities

Need affordable access to Smart City services, digital literacy programs, and economic opportunities. Specific issues include:

- Limited access to affordable housing and basic services
- High costs of internet and mobile services
- Limited economic opportunities and job insecurity

Rural-Urban Migrants

Require integration and inclusion in Smart City planning, with access to services and opportunities. Specific issues include:

- Limited access to housing and basic services
- Social exclusion and marginalization
- Limited economic opportunities and job insecurity

Other marginalized groups, such as the elderly and youth, also face unique challenges in accessing Smart City benefits.

5.9.5 Strategies for Inclusive Smart City Development

Participatory Urban Planning

Engage marginalized groups in decision-making processes to ensure their needs are adequately addressed. Strategies include:

- Community engagement and public consultations
- Inclusive zoning and land-use policies
- Participatory budgeting and resource allocation

5.9.6 Human-Centered Design

Design Smart City initiatives that prioritize the needs and experiences of all citizens. Key strategies include:

- User-centered design approaches
- Accessibility and usability testing
- Feedback mechanisms and evaluation processes

5.9.7 Digital Literacy and Access Programs

Provide training and access to technology for marginalized groups. Strategies include:

- Digital literacy programs and training
- Affordable internet and mobile services
- Public access points for technology and information

5.9.8 Affordable and Accessible Infrastructure

Ensure that Smart City infrastructure is accessible, affordable, and inclusive for all citizens. Key strategies include:

- Accessible pedestrian crossings and public spaces
- Affordable housing and basic services
- Public transportation systems and mobility services

5.9.9 Policy and Regulatory Frameworks

Develop policies and regulations that promote inclusivity, accessibility, and affordability in Smart City development.

Strategies include:

- Inclusive zoning and land-use policies
- Accessibility standards and guidelines
- Public participation and engagement processes

5.10 Submissions for Smart Cities

The widespread use of intelligent applications in recent years has significantly transformed the way we work and live. Emerging technologies, such as big data, artificial intelligence, and the Internet of Things (IoT), are driving this transformation. To address complex challenges and improve the quality of life for individuals and communities, these technologies have been applied across a variety of sectors, including healthcare, government, the environment, transportation, energy, safety and security, infrastructure, and education. Furthermore, the goal is to develop a framework for understanding the types of Smart City applications being used in different sectors and their broader societal impacts (Alahi et al., 2023).

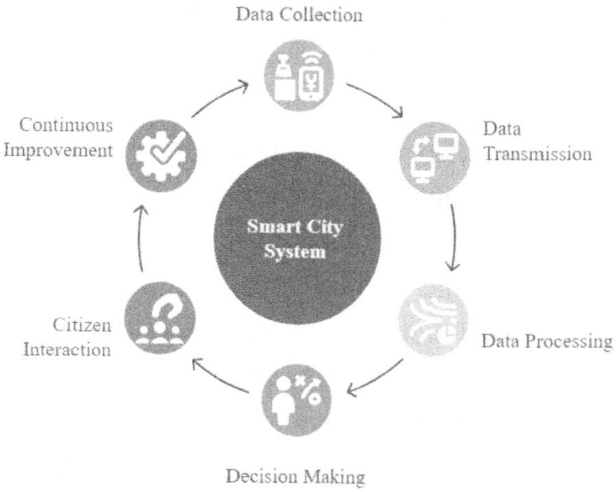

Figure 5.6: Cycle Network of Smart City System

Figure 5.6 shows that the Cycle Network of the Smart City System illustrates the cyclical and interconnected nature of Smart City systems. This system-thinking model highlights how feedback loops between data collection, policymaking, service delivery, and citizen engagement ensure adaptability. Cities that endure are those that continuously learn and self-correct, qualities embedded in smart systems design.

5.11 Conceptualizing Smart Cities Beyond Technology

Smart Cities are urban areas that strategically leverage information and communication technologies (ICTs) to enhance service delivery efficiency, promote sustainable development, and strengthen governance systems (Alahi et al., 2023). There is a critical need for long-term planning, resilience, and adaptability in the context of developing nations, as exemplified by cities such as Lagos and Jakarta.

Kabul, Nairobi, and São Paulo, grappling with complex urban challenges such as traffic congestion, poor waste management, unreliable power supply, water scarcity, and insecurity, can leverage the Smart City model as a transformative solution. Through the implementation of digital technologies, including real-time traffic management, intelligent energy and water meters, digital public opinion platforms, and surveillance security systems, these urban centers can better manage scarce resources and enhance urban living conditions.

ICTs provide the technological foundation of Smart Cities, but Smart Cities are not just about technology. The lack of a structured governance framework to coordinate digital transformation, maintain standards, and provide sectoral coordination can undermine innovation. Policies must govern the ethical use of data, ensure privacy, and standardize transparent technical integration. Furthermore, institutions must be equipped with the technological capacity, infrastructure, and financial resources to implement and sustain such innovations over the long term (Li, Feng, Ran, Su, Liu, Huang, & Guo, 2023). In Africa, where institutional capacity is often constrained by corruption, policy incoherence, and poor inter-agency coordination, the risk of Smart City initiatives becoming fragmented or underutilized is high. Therefore, the application of Smart City frameworks requires more than technology; it requires visionary leadership, inclusive policy regimes, and institutionally empowered capabilities that can translate digital potential into tangible development outcomes for urban residents.

5.12 The Backbone of Smart City Development

Governance refers to the processes, structures, institutions, and norms that influence the exercise of power, decision-making, stakeholder participation, and accountability of public officials and institutions. Governance encompasses formal mechanisms, such as laws, regulations, and administrative processes, as well as informal practices that shape the behavior of actors in society. In the context of Smart Cities, governance is not merely about controlling technology deployment; it also involves building a transparent, participatory, and inclusive system for the design, deployment, and operation of digital solutions.

Effective governance in Smart Cities is essential to ensure that technological innovations benefit all citizens rather than exacerbate marginalization or exclude vulnerable groups. It ensures that infrastructure development keeps pace with citizens' needs, that information collected from the public is managed ethically, and that systems are protected against corruption and misuse.

In Africa, where weak governance, low institutional capacity, limited transparency, decentralized policy implementation, and corruption are common, good governance is even more critical. The success of Smart City initiatives, such as the Urban City or the Smart City Initiative, does not solely depend on the level of investment in technology, but also on whether the governance framework facilitates inter-agency collaboration, citizen participation, and regulatory compliance. Moreover, inclusive governance ensures that Smart City planning accounts for the realities of a large informal economy, widespread digital illiteracy, and urban poverty.

Citizen participation through open data platforms, town hall meetings, and inclusive feedback mechanisms plays a critical role in building trust and ensuring that Smart City initiatives remain people-centered, equitable, and responsive to community needs.

Without such participatory governance structures, Smart City projects risk becoming top-down, technocratic initiatives that primarily serve elite interests while remaining disconnected from the daily realities of most urban residents. Effective governance in this context is not only about oversight but also about facilitating inclusive decision-making, accountability, and social justice (Luque-Ayala & Marvin, 2015; Ramirez Lopez & Grijalba Castro, 2020).

Smart City governance requires the proactive engagement of diverse stakeholders, including government agencies, the private sector, civil society, and citizens. Participatory and inclusive governance enables the co-production of city services that are more responsive to the needs of diverse communities. Open governance arrangements promote trust, which is essential when cities collect and utilize vast quantities of personal data. However, leadership is one of the most critical elements in governing Smart City innovations.

A vision backed by prioritized strategies aligns actors around a common agenda. City leaders are not only responsible for advocating smart innovations but also for ensuring that these initiatives are consistent with broader developmental goals, such as equity, inclusion, and sustainability (Angelidou, 2017).

5.13 Governance Models and Inter-Governmental Relations

The design of governance models can significantly influence the success of Smart City projects. Centralized governance can provide uniform standards and coordinated investments, whereas decentralized models are often better suited to address local needs. Very often, success depends on effective intergovernmental relations that align national policies with city-level programs (Meijer & Bolívar, 2016).

5.14 Policy as a Catalyst for Smart City Success

Policy plays a pivotal role in determining the direction, scope, and success of Smart City projects. It encompasses the guidelines, laws, regulations, and strategic frameworks that govern how technologies are implemented, managed, and scaled across urban systems. Policy in Smart Cities is effective because it serves as a bridge between vision and implementation, providing clarity, consistency, and direction for both public and private stakeholders.

Policies specify how information is collected and managed, how infrastructure is developed, how funding for digital services is allocated, and how citizens are protected in a digital society. In low-income countries, the absence of coherent, forward-looking urban policies has been one of the major barriers to Smart City development.

Although efforts such as the National Digital Economy Policy and Strategy (2020–2030) and the Low-Income Country Smart City Framework have been made to establish a platform for digital transformation, their implementation has sometimes been fragmented, inconsistent, and insufficiently monitored.

Most policies in cities still operate in silos, and very few ministries, departments, and agencies coordinate effectively at the federal, state, and local levels. Smart City development requires adaptive, inclusive, and technology-driven policy frameworks. Such policies should promote public-private partnerships (PPPs), incentivize innovation, and ensure interoperability of systems across sectors such as transport, energy, housing, and security. Furthermore, policies must address key contextual challenges, including closing the digital divide, financing infrastructure in underserved areas, enhancing digital literacy, and establishing city financing models that are fit for purpose.

Smart City policy must also be inclusive, engaging local governments, civil society, academia, and the private sector in both the formulation and implementation of policy. An appropriately framed and properly executed policy environment not only provides the legal and operational framework for Smart City initiatives but also ensures that they are sustainable, equitable, and effective. In the absence of such policies, Smart City initiatives risk becoming stalled, politicized, or unable to meet the demands of rapidly growing cities.

5.15 A Call to Rise: Innovating for Context and Impact

Gap Between Cutting-Edge Innovation and Practical Application in Smart Cities: Demonstrating Feasibility Without Extravagance

In the pursuit of Smart Cities, the challenge lies not merely in adopting the latest technologies but in ensuring their practical application actually enhances urban living without unnecessary complexity or cost. This approach emphasizes the importance of demonstrating feasibility through practical, scalable solutions that prioritize functionality over novelty.

Smart Cities integrate digital technologies into urban infrastructure to improve residents' quality of life, enhance sustainability, and streamline urban management. These cities utilize data, sensors, and interconnected systems to optimize services such as transportation, energy, healthcare, and public safety. However, the rapid pace of technological advancement often creates pressure to adopt cutting-edge innovations without fully considering their practical implications. While innovative technologies offer exciting possibilities, their real-world application must be carefully evaluated to ensure they effectively address actual urban challenges.

5.16 The Urgency of a New Urban Paradigm

The rapid pace of urbanization worldwide necessitates a reimagined framework for urban development—one that prioritizes inclusivity, sustainability, and resilience. Traditional urban planning approaches often fall short in addressing the multifaceted, context-specific challenges faced by cities in the Global South. In response, policymakers and urban stakeholders must embrace innovative strategies that are locally adapted, combining technological integration with community participation and environmentally sustainable practices (UN-Habitat, 2020).

Smart City initiatives should not be applied universally but should instead be tailored to reflect the socio-economic and ecological realities of each locality. A notable example is Malmö, Sweden's Hyllie district, which demonstrates the effectiveness of integrated planning that incorporates renewable energy (solar, wind, and biogas) along with smart energy grids and storage systems to enhance efficiency and reduce emissions through a collaborative public-private partnership (The Guardian, 2024).

Likewise, Chennai, India, has adopted urban greening measures such as Miyawaki forests and vertical farming, promoting biodiversity and improving urban air quality. The city has also implemented intelligent transportation systems that include adaptive traffic management and GPS-tracked vehicles, which have significantly alleviated congestion (Emerald, 2024). Truly inclusive Smart Cities are built through participatory planning processes that give voice to all social groups, especially marginalized populations. Sweden's SymbioCity model is a prime example of a people-centered urban planning strategy that emphasizes integrated development and local stakeholder involvement to uplift vulnerable urban communities (Sida, 2012). In African contexts, the deployment of digital engagement platforms has enhanced civic participation, making governance processes more transparent, accountable, and responsive to citizens' needs (UNESCO, 2023).

5.17 Leveraging Technology for Smarter Governance

Technological advancements are creating transformative opportunities for improving urban governance. Emerging tools such as artificial intelligence (AI) and the Internet of Things (IoT) facilitate real-time monitoring and data-driven management of urban systems, enabling cities to allocate resources more efficiently.

A notable example is the city of Aveiro in Portugal, where an AI-integrated platform monitors traffic patterns, energy usage, and infrastructure conditions to streamline city operations and enhance residents' quality of life (Pereira et al., 2024). However, the deployment of such technologies must be guided by ethical considerations.

As highlighted in a 2024 study, it is crucial to assess the societal and ethical implications of AI in Smart City initiatives to ensure that these innovations uphold values of equity, transparency, and the public good (Smith & Rahimi, 2024).

5.18 Roadmap and Framework

Increasing internet infrastructure, providing subsidies for smart gadgets, and encouraging an innovative culture are all components of a sustainable technology adoption strategy. Local IT hardware manufacturing can save money and create jobs. Scalable pilot initiatives in education, agriculture, and health can be funded by impact investors and donor organizations. To learn from implementations and modify them appropriately, robust monitoring and evaluation mechanisms are required. No one will be left behind if a national innovation plan places a high priority on inclusivity, including gender equity, rural access, and support for people with disabilities. Promoting diaspora involvement can also strengthen local ecosystems by bringing in capital and expertise.

Emerging technologies have enormous potential to transform low-resource environments in Nigeria, despite significant obstacles. Nigeria can leverage AI, IoT, and other technologies to promote equitable and sustainable development by adapting innovations to local needs and investing in infrastructure, education, and policy reform. These tools have the potential to be instruments of empowerment that lift millions out of poverty and advance Nigeria into the digital future with careful planning and inclusive execution (Haidine et al., 2016; Macnaghten et al., 2019; Stimmel, 2015).

5.19 Smart for All: Prioritizing Inclusivity in Urban Transformation

Inclusivity, based on the aforementioned factors, is essential for accelerating development through urbanization. However, inclusivity must be achieved despite prevailing poor local government planning capabilities and limited financial resources, particularly in developing countries. Such capacity constraints can potentially undermine the effectiveness of consultative planning activities unless adequately addressed (UCLG Afrique - Cities Alliance, 2015). The physical, social, and economic landscapes of cities are continuously shaped by the ongoing process of urban development.

To build cities that are fair, accessible, and responsive to the diverse needs of all residents, there has been an increasing focus on inclusivity in urban development. In recent decades, there has been a growing emphasis on inclusive urban planning to create communities that are equitable, accessible, and considerate of the varied needs of all citizens. Inclusive urban transformation is not merely a hope but a necessity for cities to become just, fair, and sustainable. By learning from the best international practices and implementing inclusive strategies, cities can provide every citizen with the opportunity to thrive. While challenges remain, the world is increasingly urbanizing, and by 2030, the urbanization rate is projected to reach 60%. Since over 60% of the world's gross domestic product is generated in cities and metropolitan regions, these areas are the primary drivers of economic growth.

They also account for over 60% of global resource consumption and 70% of total carbon emissions. The promise of inclusivity can create cities that reflect the diversity and aspirations of their people. Naqvi, Shah, and Abbas (2020) establish the connection between increasing urbanization and broad growth as one that generates equitable opportunities and ensures access to these opportunities that is, development that allows the public to participate and translates into individual economic advancement (Lemaire & Kerr, 2017; Samouylov et al., 2019; Pereira et al., 2018).

5.20 Comparative Gap in developing countries on low-income countries and Smart City development

Including examples from other developing nations in the discussion on Smart City development and sustainability provides valuable comparative insights that significantly enhance the relevance and depth of this chapter. While Nigeria remains the primary focus, integrating diverse and contextually grounded experiences from peer countries enriches understanding of shared challenges and innovative responses among low- and lower-middle-income countries.

Such comparative perspectives not only strengthen analytical rigor but also highlight transferable lessons and adaptable strategies for inclusive and sustainable urban transformation. First, socioeconomic inequality, widespread informal economies, institutional fragility, and inadequate infrastructure are common challenges confronting Smart City initiatives in developing countries. This chapter demonstrates how locally adapted applications of smart technologies supported by ethical data practices and context-sensitive governance frameworks can produce practical and scalable solutions even within resource-constrained environments.

By drawing on successful and ongoing initiatives in countries such as Kenya, Rwanda, Colombia, and India, the analysis highlights how innovation can be shaped to local realities rather than imposed through imported models. This comparative perspective enhances the accessibility and cross-regional relevance of the discussion, enabling readers from diverse contexts to recognize parallels with their own urban environments.

Second, the inclusion of examples from other developing regions challenges dominant narratives that frame Smart Cities as a predominantly high-income, technologically intensive agenda. Instead, it repositions Smart City development as a flexible, inclusive, and adaptive process that can be meaningfully pursued in low- and lower-middle-income settings when aligned with local needs, capacities, and governance structures.

Instead, it draws attention to creative, affordable, and community-focused strategies that are becoming popular in the Global South, such as Bogotá's data-driven public transportation reform, Chennai's eco-smart traffic solutions, and Kigali's citizen-led ICT programs. These examples prioritize contextual intelligence over imported models, which aligns with the chapter's overarching theme of inclusive, locally based, and sustainable urban development.

Additionally, by promoting knowledge sharing, these examples can help practitioners, researchers, and policymakers in Nigeria and other nations learn from one another. Beyond isolated national experiences, they provide useful insights into what works, what doesn't, and why, fostering a more dynamic understanding of Smart City conversation.

References

Abubakar, I. R., & Aina, Y. A. (2019). The prospects and challenges of developing more inclusive, safe, resilient and sustainable cities in Nigeria. *Land use policy*, *87*, 104105.

Akande, A., Cabral, P., Gomes, P., & Casteleyn, S. (2019). The Lisbon ranking for smart sustainable cities in Europe. *Sustainable cities and society*, *44*, 475-487.

Alahi, M. E. E., Sukkuea, A., Tina, F. W., Nag, A., Kurdthongmee, W., Suwannarat, K., & Mukhopadhyay, S. C. (2023). Integration of IoT-enabled technologies and artificial intelligence (AI) for Smart City scenario: recent advancements and future trends. *Sensors*, *23*(11), 5206.

Albino, V., Berardi, U., & Dangelico, R. M. (2015). Smart Cities: Definitions, dimensions, performance, and initiatives. *Journal of urban technology*, *22*(1), 3-21.

Al-Kodmany, K. (2018). *The vertical city: a sustainable development model*. WIT press.

Allam, Z., & Dhunny, Z. A. (2019). On big data, artificial intelligence and smart cities. *Cities*, *89*, 80-91.

Alsaid, L. A. Z. A. (2021). Performance measurement in Smart City governance: A case study of an Egyptian city council. *Journal of Accounting in Emerging Economies*, *11*(3), 395-430.

Angelidou, M. (2017). The role of smart city characteristics in the plans of fifteen cities. *Journal of Urban Technology, 24*(4), 3–28. https://doi.org/10.1080/10630732.2017.1348880.

Angelidou, M. (2015). Smart Cities: A conjuncture of four forces. *Cities*, *47*, 95-106.

Ateş, M. (2024). The Relationship between Smart Cities, Urban Resilience and Sustainability: Implications for Urban Planning. *Journal of Planning Architecture and Design*, *2*(1-2), 1-19.

Azevedo Guedes, A. L., Carvalho Alvarenga, J., Dos Santos Sgarbi Goulart, M., Rodriguez y Rodriguez, M. V., & Pereira Soares, C. A. (2018). Smart cities: The main drivers for increasing the intelligence of cities. *Sustainability, 10*(9), 3121.

Batty, M., Axhausen, K. W., Giannotti, F., Pozdnoukhov, A., Bazzani, A., Wachowicz, M., ... & Portugali, Y. (2012). Smart Cities of the future. *The European Physical Journal Special Topics, 214*(1), 481–518. https://doi.org/10.1140/epjst/e2012-01703-3

Belsky, E. S. (2012). Planning for Inclusive and Sustainable Urban Development. *State of the World 2012: Moving Toward Sustainable Prosperity*, 38-52.

Bibri, S. E. (2018). The IoT for smart sustainable cities of the future: An analytical framework for sensor-based big data applications for environmental sustainability. *Sustainable cities and society*, *38*, 230-253.

Bibri, S. E. (2019). On the sustainability of smart and smarter cities in the era of big data: an interdisciplinary and transdisciplinary literature review. *Journal of Big Data*, *6*(1), 25.

Caragliu, A., Del Bo, C., & Nijkamp, P. (2011). Smart Cities in Europe. *Journal of Urban Technology, 18*(2), 65–82. https://doi.org/10.1080/10630732.2011.601117

Cartwright, A., Palmer, I., Taylor, A., Pieterse, E., Parnell, S., & Colen brander, S. (2018). Developing prosperous and inclusive cities in Africa-National Urban Policies to the rescue. *London and Washington DC: Coalition for Urban Transitions*.

Castelnovo, W., Misuraca, G., & Savoldelli, A. (2016). Smart Cities governance: The need for a holistic approach to assessing urban participatory policy making. *Social Science Computer Review*, *34*(6), 724-739.

Chong, W. Y. (2021). Smart governance and Smart Cities: A review of the Singapore experience. *Cities, 110*, 103083. https://doi.org/10.1016/j.cities.2020.103083

Colding, J., Nilsson, C., & Sjöberg, S. (2024). Smart Cities for all? Bridging digital divides socially sustainable and inclusive cities. *Smart Cities*, *7*(3), 1044-1059.

Daramola, O. J. (2024). Building Sustainable and Inclusive Smart Cities. Researchgate.net

Dierwechter, Y. (2017). Urban sustainability through smart growth. *Cham: Springer. Doi*, *10*, 978-3.

Domorenok, E., Graziano, P., & Polverari, L. (2021). Introduction: Policy integration and institutional capacity: Theoretical, conceptual and empirical challenges. *Policy and Society*, *40*(1), 1-18.

Emerald Publishing. (2024). *Urban Mobility and Green Innovation in Indian Smart Cities.*

Glasmeier, A., & Christopherson, S. (2015). Thinking about smart cities. *Cambridge Journal of Regions, Economy and Society*, *8*(1), 3-12.

Glasmeier, A. K., & Nebiolo, M. (2016). Thinking about smart cities: The travels of a policy idea that promises a great deal, but so far has delivered modest results. *Sustainability*, *8*(11), 1122.

Goodman, E. P., & Powles, J. (2019). Urbanism under Google: Lessons from Sidewalk Toronto. *Fordham Law Review, 88*(2), 457–498.

Greera, C. (2022, November). Prioritizing Inclusion in Urban Development. In *2022 IEEE International Symposium on Technology and Society (ISTAS)* (Vol. 1, pp. 1-4). IEEE.

Habibzadeh, H., Nussbaum, B. H., Anjomshoa, F., Kantarci, B., & Soyata, T. (2019). A survey on cybersecurity, data privacy, and policy issues in cyber-physical system deployments in Smart Cities. *Sustainable Cities and Society*, *50*, 101660.

Haidine, A., El Hassani, S., Aqqal, A., & El Hannani, A. (2016). The role of communication technologies in building future Smart Cities. In *Smart Cities technologies*. Intech Open.

Healey, P., Cars, G., Madanipour, A., & De Magalhaes, C. (2017). Transforming governance, institutionalist analysis and institutional capacity. In *Urban governance, institutional capacity and social milieux* (pp. 6-28). Routledge.

Herrschel, T. (2013). Competitiveness and sustainability: can 'Smart City regionalism' square the circle?. *Urban Studies*, *50*(11), 2332-2348.

Ismagiloiva, E., Hughes, L., Rana, N., & Dwivedi, Y. (2019). Role of Smart Cities in creating sustainable cities and communities: A systematic literature review. In *International Working Conference on Transfer and Diffusion of IT* (pp. 311-324). Springer, Cham.

Janssen, M., & Estevez, E. (2013). Lean government and platform-based governance—Doing more with less. *Government Information Quarterly, 30*(S1), S1–S8. https://doi.org/10.1016/j.giq.2012.11.003

Kesar, P., & Ache, P. M. (2024). Past, present and future of the Smart City in India: An institutional perspective. *Cities*, *146*, 104775.

Kirimtat, A., Krejcar, O., Kertesz, A., & Tasgetiren, M. F. (2020). Future trends and current state of Smart City concepts: A survey. *IEEE access*, *8*, 86448-86467.

Kitchin, R. (2014). The real-time city? Big data and smart urbanism. *GeoJournal, 79*(1), 1–14. https://doi.org/10.1007/s10708-013-9516-8

Kostakos, V., O'Neill, E., Little, L., & Sillence, E. (2005). The social implications of emerging technologies. *Interacting with computers*, *17*(5), 475-483.

Kummitha, R. K. R., & Crutzen, N. (2017). How do we understand Smart Cities? An evolutionary perspective. *Cities*, *67*, 43-52.

Latifi, G. R., Monfared, M. P., & Khojasteh, H. A. (2022). Gamification and citizen motivation and vitality in Smart Cities: A qualitative meta-analysis study. *GeoJournal*, *87*(2), 1217-1230.

Lawton, A., Van Der Wal, Z., & Huberts, L. (2015). Ethics in Public policy and management. *A Global Research Companion. Milton Park: Routledge. [Google Scholar]*.

Lee, J., Babcock, J., Pham, T. S., Bui, T. H., & Kang, M. (2023). Smart city as a social transition towards inclusive development through technology: a tale of four Smart Cities. *International Journal of Urban Sciences*, *27*(sup1), 75-100.

Lee, J. H., Hancock, M. G., & Hu, M. C. (2014). Towards an effective framework for building Smart Cities: Lessons from Seoul and San Francisco. *Technological Forecasting and Social Change*, *89*, 80-99.

Lee, S., & Park, I. K. (2025). Key Themes and Conceptual Frameworks for a Smart-Inclusive City: A Scoping Review. *Journal of Planning Literature*, 08854122241312909.

Lemaire, X., & Kerr, D. (2017). Inclusive urban planning–promoting equality and inclusivity in urban planning practices.

Li, J., Xu, M., Xiang, L., Chen, D., Zhuang, W., Yin, X., & Li, Z. (2024). Foundation models in smart agriculture: Basics, opportunities, and challenges. *Computers and Electronics in Agriculture*, *222*, 109032.

Lowery-Gionta, E. G., Stritzel, H., Matson, L. M., Bucher, M. A., An, X., Ji, Y., ... & Adler, A. B. (2025). Initial development of the Stress Monitoring and Response Tool (SMART): A holistic measure of stress following trauma. *PloS one*, *20*(6), e0321939.

Luque-Ayala, A., & Marvin, S. (2015). Developing a critical understanding of smart urbanism? Urban Studies, 52(12), 2105–2116.

Lyu, W., Wang, T., Hou, R., & Liu, J. (2023). Going green and profitable: The impact of smart manufacturing on Chinese enterprises. *Computers & Industrial Engineering*, *181*, 109324

Lyu, Y., Lu, H., Lee, M. K., Schmitt, G., & Lim, B. Y. (2022). IF-City: Intelligible Fair City Planning to Measure, Explain, and Mitigate Inequality. *arXiv*. https://arxiv.org/abs/2202.07349

Macke, J., Casagrande, R. M., Sarate, J. A. R., & Silva, K. A. (2018). Smart City and quality of life: Citizens' perception in a Brazilian case study. *Journal of cleaner production*, *182*, 717-726.

Macnaghten, P., Davies, S. R., & Kearnes, M. (2019). Understanding public responses to emerging technologies: a narrative approach. *Journal of Environmental Policy & Planning*, *21*(5), 504-518.

March, H., & Ribera-Fumaz, R. (2016). Smart contradictions:

The politics of making Barcelona a self-sufficient city. *European Urban and Regional Studies, 23*(4), 816–830. https://doi.org/10.1177/0969776414554488

Martin, C. J., Evans, J., & Karvonen, A. (2018). Smart and sustainable? Five tensions in the visions and practices of the smart-sustainable city in Europe and North America. *Technological Forecasting and Social Change*, *133*, 269-278.

Martin, C., Evans, J., Karvonen, A., Paskaleva, K., Yang, D., & Linjordet, T. (2019). Smart-sustainability: A new urban fix?. *Sustainable cities and society*, *45*, 640-648.

Meijer, A., & Bolívar, M. P. R. (2016). Governing the Smart City: A review of the literature on smart urban governance. *International Review of Administrative Sciences, 82*(2), 392–408. https://doi.org/10.1177/0020852314564308

Mitlin, D. (2014). Towards inclusive urban planning and practice.https://scholar.google.com/scholar?hl=en&as_sdt=0%2C5&q=Mitlin%2C+D.+%282014%29.+Towards+inclusive+urban+planning+and+practice.&btnG=#:~:text=2014%E2%80%A2-,diva%2Dportal.org,-The%20contribution%20of

Nam, T., & Pardo, T. A. (2011). Smart City as urban innovation:
Focusing on management, policy, and context. *Proceedings of the 5th International Conference on Theory and Practice of Electronic Governance*, 185–194. https://doi.org/10.1145/2072069.2072100

Naqvi, N., Rehman, S., & Islam, Z. (2020). A hyperconnected Smart City framework: Digital resources using enhanced pedagogical techniques. *Australasian Journal of Information Systems*, *24*, 1-42.

Neirotti, P., De Marco, A., Cagliano, A. C., Mangano, G., & Scorrano, F. (2014). Current trends in Smart City initiatives: Some stylised facts. *Cities*, *38*, 25-36.

Nesti, G. (2020). Defining and assessing the transformational nature of smart city governance: Insights from four European cases. *International Review of Administrative Sciences*, *86*(1), 20-

37.
OECD. (2020). *Smart Cities and inclusive growth: Building on The outcomes of the 1st OECD Roundtable on Smart Cities and Inclusive Growth*. https://www.oecd.org

Oliveira, Á., & Campolargo, M. (2015, January). From Smart Cities to human smart cities. In *2015 48th Hawaii International Conference on System Sciences* (pp. 2336-2344). IEEE.

Painter, M., & Pierre, J. (2005). Unpacking policy capacity: Issues and themes. In *Challenges to state policy capacity: Global trends and comparative perspectives* (pp. 1-18). London: Palgrave Macmillan UK.

Pereira, G. V., Parycek, P., Falco, E., & Kleinhans, R. (2018). Smart governance in the context of Smart Cities: A literature review. *Information Polity*, *23*(2), 143-162.

Pierce, P.; Ricciardi, F.; Zardini, A. Smart Cities as organizational fields: A framework for mapping Ramirez Lopez, L. J., & Grijalba Castro, A. I. (2020). Sustainability and resilience in Smart City planning: A review. *Sustainability*, *13*(1), 181.

Pereira, J., Costa, L., & Moreira, R. (2024). *AI-Driven Urban Management: A Case Study of Aveiro, Portugal*. arXiv preprint arXiv:2401.12345.

Razaghi, M.; Finger, M. (2018) Smart governance for Smart Cities. Proc. IEEE, 206, 680–689. [CrossRef]

Ramirez Lopez, L., & Grijalba Castro, A. S. (2020). Governance and citizen participation in Smart Cities: Rethinking inclusive approaches. Journal of Urban Affairs, 42(4), 567–583.

Renda, A. (2019). *Artificial Intelligence. Ethics, governance and policy challenges*. CEPS Centre for European Policy Studies.

Rotolo, D., Hicks, D., & Martin, B. R. (2015). What is an emerging technology? *Research policy*, *44*(10), 1827-1843.

Ruhlandt, R.W.S. (2018) The governance of Smart Cities: A Systematic literature review. Cities, 81, 1 [CrossRef]

Samouylov, K., Popov, E., & Semyachkov, K. (2019). Institutional support of a Smart City.

Sampson, R. J. (2017). Urban sustainability in an age of Enduring inequalities: Advancing theory and ecometrics for the 21st-century city. *Proceedings of the National Academy of Sciences, 114*(34), 8957-8962.

Shmelev, S.E.; Shmeleva, I.A. (2019). Multidimensional sustainability benchmarking for smart megacities. Cities, 91, 134–163. [CrossRef]

Silva, B.N.; Khan, M.; Han, K. (2018). Towards sustainable smart cities: A review of trends, architectures, components, and open challenges in Smart Cities. Sustain. Cities Soc., 38, 697–713. [CrossRef]

Singla, M., Gupta, D., Goyal, D., & Dang, R. K. (2024, June). Integrating Smart Farming with AI Techniques for Crop Disease Prediction. In *2024 15th International Conference on Computing Communication and Networking Technologies (ICCCNT)* (pp. 1-6). IEEE.

Singla, P., Singh, A., Gupta, A., & Garg, S. (2024). Adaptive Urban Planning: A Hybrid Framework for Balanced City Development. *arXiv*. https://arxiv.org/abs/2412.15349

Sida. (2012). *SymbioCity Approach: Towards Sustainable Urban Development*. Stockholm: Swedish International Development Cooperation Agency.

Smith, K. (2020). The Inconvenient Truth About Smart Cities. https://blogs.scientificamerican.com/observations/the-inconvenient-truth-about-smart-cities/ (accessed on 27 January 2020).

Smith, T., & Rahimi, S. (2024). *Ethical Frameworks for Artificial Intelligence in Smart Cities*. Journal of Urban Technology, 31(2), 115–132.

Stimmel, C. (2015). *Building Smart Cities*. New York, NY, USA: Auerbach Publications.

Times of India. (2025). Odisha presents housing strategy at ADB forum. *Times of India*. https://timesofindia.indiatimes.com/city/bhubaneswar/odisha-presents-housing-strategy-at-ad-forum/articleshow/121472149.cms

Townsend, A. M. (2013). *Smart Cities: big data, civic hackers, and the quest for a new utopia*. W. W. Norton & Company

The Guardian. (2024). *How Hyllie Became Sweden's Greenest District*. https://www.theguardian.com

UNESCO. (2023). Building inclusive urban futures: UNESCO and NIUA spotlight event explores data for people with disabilities. *UNESCO*. https://www.unesco.org/en/articles/building-inclusive-urban-futures-unesco-and-niua-spotlight-event-explores-data-people-disabilities

UN-Habitat. (2020). *Planning for Inclusive Urban Development*.

UN-Habitat. (2020). *World Cities Report 2020: The Value of Sustainable Urbanization*. https://unhabitat.org

UN-Habitat. (2023). Delivering inclusive cities through practical action: Launching the Global Action Report on Inclusive Design and Infrastructure in Cities. *World Urban Forum 12*. https://wuf.unhabitat.org/event/wuf12/delivering-inclusive-cities-through-practical-action-launching-global-action-report

United Nations. World Urbanization Prospects: The 2018 Revision [Key Facts]; UN: New York, NY, USA, 2018.

UNESCO. (2023). *Building Inclusive Urban Futures: Global Action Report on Inclusive Design and Infrastructure in Cities*. https://www.unesco.org

Verhulst, S. & Young, A. (2017). *Open Data in Developing Economies: Toward Building an Evidence Base on What Works and How*. The GovLab.

Watson, V. (2009). The planned city sweeps the poor away: Urban planning and 21st-century urbanization. *Progress in Planning*, 72(3), 151–193.

WEF (World Economic Forum). (2020). *The Smart City Playbook: Kigali*. https://www.weforum.org

Whitzman, C., Legacy, C., Andrew, C., Klodawsky, F., Shaw, M., & Viswanath, K. (2013). Building inclusive cities. *Women's*

Safety and the Right to the City.

Winters, J. V. (2011). Why are Smart Cities growing? Who moves who stays? J. Reg. Sci., 51, 253–270. [CrossRef]

World Bank. (2015). *Inclusive Cities Approach Paper*

Yigitcanlar, T.; Han, H.; Kamruzzaman, M.; Ioppolo, G.; Sabatini-Marques, J. (2019). The making of Smart Cities: Are Songdo, Masdar, Amsterdam, San Francisco, and Brisbane the best we could build? Land Use Policy, 88, 104187. [CrossRef]

Yigitcanlar, T.; Kamruzzaman, M.; Buys, L.; Ioppolo, G.; Sabatini-Marques, J.; da Costa, E.M.; Yun, J.J. (2018). Understanding 'Smart Cities': Intertwining development drivers with desired outcomes in a multidimensional framework. Cities, 81, 145–160. [CrossRef]

Zuboff, S. (2019). *The age of surveillance capitalism: The fight for a human future at the new frontier of power.* PublicAffairs.

ABOUT THE AUTHORS

Professor Dr. Seyed Mohammadreza Ghadiri
Professor and Chancellor at Swiss Innovation and Management Institute, Switzerland

BIODATA

Professor Dr. Seyed Mohammadreza Ghadiri is a distinguished civil engineer with a Ph.D. in Transportation Engineering, specializing in Intelligent Transportation Systems, and a Master's degree in Structural Engineering. His research interests include logistics, smart cities, smart mobility, and sustainable development, reflecting a strong engagement with emerging global challenges and innovations. He currently serves as Professor and Chancellor of the Swiss Information and Management Institute (SIMI), Switzerland. Previously, he held senior academic leadership positions, including Dean of the School of Transportation and Logistics and Head of the Graduate School at the Malaysia University of Science and Technology. Internationally recognized for his expertise, Professor Ghadiri is a frequent keynote speaker and guest professor. He has delivered lectures, research presentations, and professional training across numerous countries, including China, Iran, Spain, Poland, France, Vietnam, India, Myanmar, Nigeria, and Switzerland.

Associate Professor Dr. Yekini Ojenya Salawu
Adjunct Associate Professor at Virtual World Institute in Indiana, USA

BIODATA

Dr. Yekini Ojenya Salawu is an Adjunct Associate Professor at the Virtual World Institute, Indiana, USA, and Co-Founder and Vice President of Translog Innovation Academy, Nigeria. He is a seasoned scholar and industry professional with over 25 years of combined academic and industry experience. Dr. Salawu holds a PhD in Transportation and Logistics from the Malaysia University of Science and Technology (MUST), alongside an MBA in Supply Chain Management, a Master of Transport and Logistics, a Postgraduate Diploma in Logistics and Supply Chain Management, and a BSc in Mathematics from Ahmadu Bello University, Nigeria. His professional experience spans manufacturing, international trade, customs brokerage, and trade logistics, and he currently works with the Nigerian Security Printing and Minting Plc. His research interests include smart cities, logistics, and sustainable supply chains. Dr. Salawu is a published author, reviewer, conference speaker, and a Fellow of leading professional institutions, including the Chartered Institute of Logistics and Transport.

Dr. Nguyen Thi Thanh Huong
Institute for European and Americas Studies

BIODATA

Dr. Nguyen Thị Thanh Hương is Director of International Cooperation at the Institute for European and Americas Studies, Vietnam Academy of the Social Sciences, a leading government think tank. She is a distinguished policy advisor and certified scientist with over 15 years of experience spanning real estate, international cooperation, UNDP projects, and business training. Her research interests include international and Vietnamese real estate, smart cities, international relations, and business management. Dr. Hương holds a Doctorate in Business Administration from Horizons University, France, specializing in real estate management, and an MBA from La Trobe University, Australia, ranked among the world's top MBA programs. She is a member of the International Panel of Sustainability Experts, honoured by Global Ambassadors of Sustainability and Global Green Chamber of Commerce. An accomplished author and reviewer, she has published books and scientific papers, including Horizons of Housing: Uncovering Contemporary Global Real Estate & Future Trends. She has also organized and moderated major international conferences, significantly fostering Vietnam's engagement with global partners and contributing to foreign policy dialogue.

Associate Professor Dr. Fatai A. Yakeen
Head of Department of Transport Management and Operations, School of Transport and Logistics, Lagos State University, Nigeria.
President of Translog Innovation Academy, Nigeria
Regional Director for Africa at the Virtual World Institute, United States

BIODATA

Associate Professor Dr. Fatai A. Yakeen is Head of the Department of Transport Management and Operations at the School of Transport and Logistics, Lagos State University (LASU), Nigeria. He holds a Bachelor's degree in economics and an MBA (Management) from LASU, a Master's degree in Transport Economics from the University of Leeds, United Kingdom, and a PhD in Transport and Logistics from the Malaysia University of Science and Technology (MUST). Dr. Yakeen has contributed extensively to national and international research and consultancy projects. His scholarly output includes peer-reviewed journal articles, co-authored academic books, and conference presentations across Africa, Europe, and Asia. He is a Fellow of the Chartered Institute of Logistics and Transport (CILT) Nigeria and the Chartered Institute of Transport Administration (CIOTA) Nigeria. Beyond academia, he serves as President of Translog Innovation Academy, Nigeria, and Regional Director for Africa at the Virtual World Institute, USA.

Associate Professor Dr. Giwa O. Mojeed
Adjunct Associate Professor at the Virtual World Institute, USA
Lecturer at the School of Transport and Logistics, Lagos State University, Nigeria

BIODATA

Dr. Mojeed Olayiwola Giwa is an Adjunct Associate Professor at the Virtual World Institute, USA, and a Lecturer in the Department of Transport Technology, School of Transport and Logistics, Lagos State University, Nigeria. He holds a PhD in Transportation and Logistics from the Malaysia University of Science and Technology and is a recognized expert in rail transport and transport modelling. With over a decade of academic experience, Dr. Giwa has published more than 30 articles in peer-reviewed, indexed journals and has authored or co-authored five academic books. Beyond teaching and research, he is Co-Founder and Vice President (Academic) of Translog Innovation Academy, Nigeria, where he actively advances research, innovation, and capacity building in transport and logistics.

Mr. Attah Kumah Mensah
Regional Head of the Driver and Vehicle Licensing Authority (DVLA), Central Region, Cape Coast-Ghana
PhD Student in Transportation & Logistics at Malaysia University of Science and Technology

BIODATA

Mr. Attah Kumah Mensah is the Regional Head of the Driver and Vehicle Licensing Authority (DVLA), Central Region, Cape Coast, Ghana, where he oversees operations, human and financial resources, and policy implementation across the region, including the Winneba and Dunkwa-On-Offin district offices. He was a key member of the committee that reviewed Ghana's Highway Code in 2023, modernizing the 1974 Law to enhance road safety, enforcement, and compliance. Mr. Mensah has participated in several international capacity-building programs, including World Bank–sponsored study tours to Kigali, Rwanda, and Boston, USA, focused on electric mobility solutions, as well as technical visits to OPUS Inspection Inc. and U.S. motor vehicle agencies. He holds an MSc in Logistics Management from Coventry University, UK, and is a Certified Professional in Logistics Management. He is currently a PhD candidate in Transportation and Logistics at the Malaysia University of Science and Technology and a co-author of "Smart City Transformation: Navigating Challenges, Building the Future".

SPOT
PUBLISHER

Made in the USA
Monee, IL
05 November 2026

67814332R00134